ANALOG OPTICAL LINKS

Analog Optical Links presents the basis for the design of analog links. Following an introductory chapter, there is a chapter devoted to the development of the small signal models for common electro-optical components used in both direct and external modulation. However, this is not a device book, so the theory of their operation is discussed only insofar as it is helpful in understanding the small signal models that result. These device models are then combined to form a complete link. With these analytical tools in place, a chapter is devoted to examining in detail each of the four primary link parameters: gain, bandwidth, noise figure and dynamic range. Of particular interest is the inter-relation between device and link parameters. A final chapter explores some of the tradeoffs among the primary link parameters.

CHARLES H. COX, III Sc.D., is one of the pioneers of the field that is now generally referred to as analog or RF photonics. In recognition of this work he was elected a Fellow of the IEEE for his contributions to the analysis, design and implementation of analog optical links. Dr. Cox is President and CEO of Photonic Systems Inc., which he founded in 1998. He holds six US patents, has given 45 invited talks on photonics and has published over 70 papers on his research in the field of phototonics.

ANALOG OPTICAL LINKS

Theory and Practice

CHARLES H. COX, III

CAMBRIDGE
UNIVERSITY PRESS

CAMBRIDGE UNIVERSITY PRESS
Cambridge, New York, Melbourne, Madrid, Cape Town, Singapore, São Paulo

Cambridge University Press
The Edinburgh Building, Cambridge CB2 2RU, UK

Published in the United States of America by Cambridge University Press, New York

www.cambridge.org
Information on this title: www.cambridge.org/9780521621632

© Cambridge University Press 2004

First published 2004
This digitally printed first paperback version 2006

A catalogue record for this publication is available from the British Library

Library of Congress Cataloguing in Publication data
Cox, Charles Howard.
Analog optical links : theory and practice / Charles H. Cox, III.
p.   cm.
Includes bibliographical references and index.
ISBN 0 521 62163 1
1. Optical communications.   2. Fiber optics.   I. Title.
TK5103.592.F52C69   2004
621.382′7 – dc22   2003055596

ISBN-13 978-0-521-62163-2 hardback
ISBN-10 0-521-62163-1 hardback

ISBN-13 978-0-521-02778-6 paperback
ISBN-10 0-521-02778-0 paperback

To Carol
and
to the memory of Charles H. Cox, Jr. and John A. Hutcheson,
whose combined influences on me defy measure or acknowledgement

Contents

Preface   *page* xi

1   Introduction   1
   1.1   Background   1
   1.2   Applications overview   8
      1.2.1   Transmit optical links   8
      1.2.2   Distribution optical links   9
      1.2.3   Receive optical links   11
   1.3   Optical fibers   12
   References   17

2   Link components and their small-signal electro-optic models   19
   2.1   Introduction   19
      2.1.1   Notation   20
   2.2   Modulation devices   20
      2.2.1   Direct modulation   20
      2.2.2   External modulation   34
   2.3   Photodetectors   49
   Appendix 2.1   Steady state (dc) rate equation model for
                 diode lasers   54
   Appendix 2.2   Absorption coefficient of an electro-absorption
                 modulator   63
   References   63

3   Low frequency, short length link models   69
   3.1   Introduction   69
   3.2   Small-signal intrinsic gain   70
      3.2.1   Direct modulation   72
      3.2.2   External modulation   74
   3.3   Scaling of intrinsic gain   75

	3.3.1	Optical power	75
	3.3.2	Wavelength	79
	3.3.3	Modulation slope efficiency and photodetector responsivity	81
3.4		Large signal intrinsic gain	82
Appendix 3.1		External modulation links and the Manley–Rowe equations	87
References			88

4 Frequency response of links 91

4.1		Introduction	91
4.2		Frequency response of modulation and photodetection devices	93
	4.2.1	Diode lasers	93
	4.2.2	External modulators	98
	4.2.3	Photodetectors	105
4.3		Passive impedance matching to modulation and photodetection devices	110
	4.3.1	PIN photodiode	112
	4.3.2	Diode laser	117
	4.3.3	Mach–Zehnder modulator	129
4.4		Bode–Fano limit	138
	4.4.1	Lossy impedance matching	139
	4.4.2	Lossless impedance matching	142
Appendix 4.1		Small signal modulation rate equation model for diode lasers	152
References			156

5 Noise in links 159

5.1		Introduction	159
5.2		Noise models and measures	160
	5.2.1	Noise sources	160
	5.2.2	Noise figure	167
5.3		Link model with noise sources	168
	5.3.1	General link noise model	168
	5.3.2	RIN-dominated link	169
	5.3.3	Shot-noise-dominated link	173
5.4		Scaling of noise figure	178
	5.4.1	Impedance matching	179
	5.4.2	Device slope efficiency	180
	5.4.3	Average optical power	182
5.5		Limits on noise figure	185
	5.5.1	Lossless passive match limit	185

	5.5.2	Passive attenuation limit	187
	5.5.3	General passive match limit	189
	Appendix 5.1	Minimum noise figure of active and passive networks	196
	References		199
6	Distortion in links		201
	6.1	Introduction	201
	6.2	Distortion models and measures	202
	6.2.1	Power series distortion model	202
	6.2.2	Measures of distortion	205
	6.3	Distortion of common electro-optic devices	217
	6.3.1	Diode laser	217
	6.3.2	Mach–Zehnder modulator	222
	6.3.3	Directional coupler modulator	225
	6.3.4	Electro-absorption modulator	227
	6.3.5	Photodiode	228
	6.4	Methods for reducing distortion	232
	6.4.1	Primarily electronic methods	233
	6.4.2	Primarily optical methods	240
	Appendix 6.1	Non-linear distortion rate equation model for diode lasers	249
	References		259
7	Link design tradeoffs		263
	7.1	Introduction	263
	7.2	Tradeoffs among intrinsic link parameters	263
	7.2.1	Direct modulation	263
	7.2.2	External modulation	268
	7.2.3	SNR vs. noise limits and tradeoffs	273
	7.3	Tradeoffs between intrinsic link and link with amplifiers	277
	7.3.1	Amplifiers and link gain	277
	7.3.2	Amplifiers and link frequency response	278
	7.3.3	Amplifiers and link noise figure	278
	7.3.4	Amplifiers and link IM-free dynamic range	279
	References		284
Index			285

Preface

In the preface I think it is better if I abandon the formality of the text and address you the reader, directly.

As I hope you will have gathered from the title, this is a book that attempts to lay out the basis for the design of analog optical links. Let me give an example that should drive home this point. It is customary in books on lasers to start with an extensive presentation based on the rate equations (do not worry at this point if you do not know what these are). In this book we also discuss lasers, but the rate equations are relegated to an appendix. Why? Because in over 15 years of link design, I have never used the rate equations to design a link! So why all the emphasis on the rate equations in other texts? Probably because they are targeted more to, or at least written by, *device designers*. The view in this book is that you are a *user of devices*, who is interested in applying them to the *design of a link*. Of course to use a device most effectively, or even to know which device to choose for a particular link design, requires some knowledge of the device, beyond its terminal behavior. To continue the laser example, it is important to know not only what the laser frequency response is, but also how it changes with bias. Hence my intent was to include sufficient information for you to *use* various electro-optic devices, but not enough information to *design* such devices.

This book is written as an introduction to the field of link design. This was an easy choice, since, to my knowledge, there are no other books exclusively covering this topic. In the early days, once the device design was complete, link "design" consisted simply of connecting a couple of the appropriate devices together with an optical fiber. However, such links always had performance that was found lacking when evaluated using any one of a number of figures of merit. The traditional approach to overcome these shortcomings was to augment the link with pre- and/or post-amplifiers. These amplifiers did improve some aspects of the performance; notably the amplifier gain could overcome the link loss. But these amplifiers introduced their own tradeoffs that complicated the task of the system designer. Further, they obscured for the device designer the impacts on link performance that improved devices would have.

Hence there emerged the need to evaluate the tradeoffs among device, link and system parameters of an *intrinsic* link, i.e. one without amplifiers. This is the best I can do to define what I mean by link design. Of course to do this I needed some sort of analytical framework. There are lots of analytical tools I could have used for this. Given my background in electrical engineering, I chose to apply the incremental, or small-signal, modeling approach that has been so successfully applied to the analysis of electronic components, such as diodes, transistors, etc.

To my surprise, the introduction of the incremental modeling approach to link design permitted design insights that are easy to overlook when you take a purely device-oriented view. For example, an early demonstration of the impact of the small-signal link design approach showed that – with proper *link* design – it was possible to eliminate high link loss, in the sense of RF out vs. RF in, without *any* change in the devices used. This is but one, albeit dramatic, example of the power of this approach. Hence, once you have worked your way through this text, you will be equipped with a systematic basis for evaluating link designs and for understanding the tradeoffs among device, link and system parameters. This is becoming increasingly important as link designers are pressed to extract the maximum performance for the minimum cost.

I have tried to write this book so that it would be accessible to three groups of readers: electrical engineers, who usually do not know much about photonics; device designers, who typically have more of a physics background that does not include much about electrical engineering; and system designers, who need a more in-depth understanding of the relationship between these areas. Take as an example Chapter 2, which covers electro-optic devices and their incremental or small-signal models. Those of you who have an electrical engineering background can skim the incremental modeling parts of this chapter, and focus more on the aspects of the electro-optic devices. Conversely, those with a device background will likely skim the device descriptions and focus more on the incremental modeling discussions. Those of you with a systems perspective may focus on the limits of link performance in terms of device parameters.

Another dimension of the accessibility space is the familiarity of the reader with the field. Those of you who are new to the field (and we need all the new blood we can in this field!) are likely to want to get the basics down – which also tend to have general applicability – before tackling the more advanced topics – which often are of interest only in specific applications. As a guide to which sections you might want to skip on a first reading, I have prepared the following table.

Introductory	Advanced/Reference
Chapter 1 – all	
Chapter 2 – all except as listed at right	Sections 2.2.1.2, 2.2.1.3, 2.2.2.2, 2.2.2.3
Chapter 3 – all	
Chapter 4 – all except as listed at right	Section 4.4
Chapter 5 – all except as listed at right	Section 5.5
Chapter 6 – all except as listed at right	Sections 6.3.3, 6.3.4, 6.4
Chapter 7 – all	

Those new to the field would also probably want to skip all the appendices on a first reading.

As for background, I have tried to make this book as self-sufficient as possible, while keeping it to a reasonable length. Where more background was needed than was feasible to include, I have given you references that can provide the needed information. I would think that if you have the background equivalent to a senior level in electrical engineering, you should be able to follow the majority of material in this book. Those with a background equivalent to a senior level in physics should also be able to follow most of the text, with perhaps the exception of the frequency response models of Chapter 4.

I would like to begin the acknowledgements by thanking all the members of the microwave photonics community. Their numerous questions over the years, not only of me but of others whom they have asked at conferences, have been invaluable in sharpening my own understanding of this material.

When this incremental modeling approach was first published, it generated some controversy, primarily because of the predictions of link RF power gain. However, there were two people who understood this approach then and have been instrumental in guiding my thinking of it over the years: hence my deep appreciation to Professors William (Bill) Bridges of the California Institute of Technology (Caltech) and Alwyn Seeds of University College London.

Several colleagues graciously agreed to read through an early draft of the entire manuscript and offered numerous helpful suggestions; thanks to Professors Bill Bridges, Caltech, Jim Roberge, MIT and Paul Yu, UCSD. I would also like to thank Professor Paul Yu who used an early draft of the manuscript in teaching his course on electro-optics at UCSD. Several other colleagues read specific chapters and offered helpful comments as well; thanks to Ed Ackerman of Photonic Systems, Chapters 5 and 7; Gary Betts of Modetek, Chapter 6; Harry Lee of MIT, rate equation appendices and Joachim Piprek of UCSB, Chapter 2. Thanks to Joelle Prince and Harold Roussell, both of Photonic Systems, for designing several of the experimental links and taking the data that are reported in this text. I appreciate the help of Ed Ackerman, who read through the entire proof copy of the manuscript, and with red pen at the ready, offered numerous suggestions. Ed also proved to be a wonderful sounding board to test presentation ideas before they were fully developed. Thanks also to John Vivilecchia now at MIT Lincoln Laboratory, for his help with early versions of some of the figures. And finally thanks to my wife Carol, for all her patience and support, as always.

It is a pleasure to acknowledge the staff at Cambridge University Press with whom it has been a delight to work; primary among them are Philip Meyler, Simon Capelin, Carol Miller and Margaret Patterson.

It seems that every time I glance through the manuscript I find another item I wish I could change. Hence I have no illusions, despite all the expert advice I have received, that the present version is "perfect" in any respect. Thus I would appreciate hearing from you with comments, suggestions and corrections. Any errors that remain are my responsibility alone.

1

Introduction

1.1 Background

Optical communication links have probably been around for more than a millennium and have been under serious technical investigation for over a century, ever since Alexander Graham Bell experimented with them in the late 1800s. However, within the last decade or so optical links have moved into the communications mainstream with the availability of low loss optical fibers. There are of course many reasons for this, but from a link design point of view, the reason for fiber's popularity is that it provides a highly efficient and flexible means for coupling the optical source to a usually distant optical detector. For example, the optical loss of a typical terrestrial 10-km free-space optical link would be at least 41 dB[1] (Cowar, 1983), whereas the loss of 10 km of optical fiber is about 3 dB at wavelengths of ~1.55 μm. To put the incredible clarity of optical fibers in perspective, if we take 0.3 dB/km as a representative loss for present optical fibers, we see that they are more transparent than clear air, which at this wavelength has an attenuation of 0.4 to 1 dB/km (Taylor and Yates, 1957).

Today the vast majority of fiber optic links are digital, for telecommunications and data networks. However, there is a growing, some might say exploding, number of applications for analog fiber optic links. In this case, the comparison is not between an optical fiber and free space but between an optical fiber and an electrical cable. Figure 1.1 shows typical cable and optical fiber losses vs. length. As can be seen, the highest loss for optical fiber is lower than even large coax for any usable frequency.

For the purposes of discussion in this book, an *optical link* will be defined as consisting of all the components required to convey an electrical signal over an optical carrier. As shown in Fig. 1.2, the most common form of an optical link can

---

[1] The decibel (dB) is defined as 10 log (ratio); in this case the ratio is that of the optical power at the photodetector to the optical power at the optical modulation source. Thus a loss of 40 dB corresponds to a power ratio of 0.0001.

1

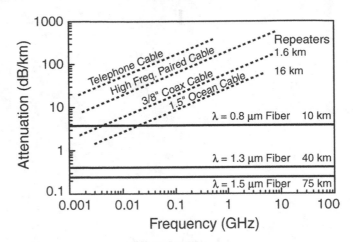

Figure 1.1 Loss versus length for representative types of electrical cables and optical fibers at three common wavelengths.

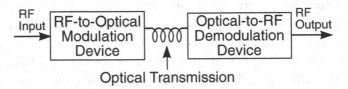

Figure 1.2 Basic components of a fiber optic link: modulation device, optical fiber and photodetection device.

be implemented with just three principal parts. At the input end is a *modulation* device, which impresses the electrical signal onto the optical carrier. An optical fiber couples the modulation device output to the input of the *photodetection* device, which recovers the electrical signal from the optical carrier.

To make the link and some of the technical issues surrounding it more concrete, consider the following example. For the modulation device we will use a diode laser and for the photodetection device a photodiode. Both of these devices will be described in detail in Chapter 2, so for now it is sufficient to know that the former converts an electrical current into a corresponding optical intensity while the latter does the reverse – it converts an optical intensity into an electrical current. We will connect these two devices optically via a length of optical fiber.

Now let us send an RF signal over this simple link. When we measure the RF signal power that we recover from the photodiode we find that we typically only get 0.1% of the RF power we used to modulate the diode laser – i.e. an RF loss of 30 dB! This raises a host of questions, among them: where did the remaining 99.9% of the RF power go; do we always have to suffer this incredible loss; what

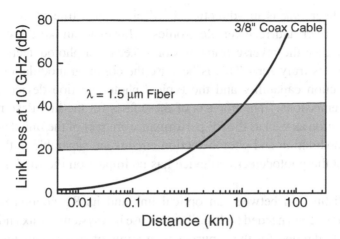

Figure 1.3 Typical loss vs. length of coax and optical fiber links operating at 10 GHz.

are the tradeoffs if we try to reduce this loss; how does this loss impact other link parameters such as the noise and distortion performance? It is the goal of this book to provide the background to answer such questions.

We can get an indication of the basis for these losses if we look at the typical loss of a link versus the length of the optical fiber between the modulation and photodetection devices. An example of this is shown in Fig. 1.3, which plots the RF loss vs. length for fiber and coaxial links operating at 10 GHz. The range of losses shown for the optical fiber link is representative of what has been reported to date. We can see that the fiber link loss increases slowly with fiber length as we would expect from the fiber optical loss data of Fig. 1.1, whereas the coax link loss increases much more quickly with length. However, note that at zero link length, the coaxial cable loss goes to zero while the optical fiber link loss does not. The zero length loss for the optical fiber link represents the combined effects of the RF/optical conversion inefficiencies of the modulation and photodetection devices. For long length links this zero length conversion loss is less important because the sum of the conversion and fiber losses is still less than the coaxial loss. But for shorter length links, where the fiber loss is negligible, the conversion loss dominates the link loss and exceeds the coaxial loss. Consequently an important aspect of link design will be understanding the reasons behind the conversion loss and developing techniques for reducing it.

In comparing an optical link with the coax or waveguide that it often replaces, there are a couple of important facts that impact link design, in addition to the loss vs. length issue we just discussed. One fact is that while the *fiber* is just as bi-directional as coax, when one includes the modulation and photodetection devices, the fiber *link* is uni-directional.[2] (This is also true of coax, when one includes the driver and receiver electronics.) However, unlike the coax case, in the fiber link case the reverse transmission – i.e. from photodetection to modulation device – is truly zero. This is because the common modulation device has no photodetection capability and the typical photodetection device cannot produce optical emissions.[3] The impacts of these facts for the link designer are that: (1) the application as well as the RF performance are part of the link design process and (2) the modulation and photodetection circuits are separable in that changing the loading at the photodetection device has no impact on the modulation device circuit.

Another distinction between an optical link and its RF counterpart is in the number of parameters needed to describe their use in a system. Coax and waveguide are completely defined for these purposes in terms of two parameters: their loss and frequency response. An optical link, which is more analogous to an active RF component – such as an amplifier – than to passive coax, typically requires at least four parameters: loss, bandwidth, noise figure and dynamic range. In terms of these four parameters, we will see that the modulation device has the greatest impact on all four parameters, with the photodetector a close second in terms of these same parameters. The fiber, especially when longer lengths are involved, can have a significant impact on loss, which as we will see in turn affects noise figure. The fiber can also indirectly affect bandwidth via its dispersion; fiber effects on dynamic range are negligible.

The emphasis in this book will be on developing the tools and techniques that will enable one to design *links* for a variety of applications, based on given device designs. This is quite different from other books where the emphasis is on designing *devices*, with secondary – at best – consideration on applying the device in a link. While the link models will be firmly rooted in the device physics, the focus here will be on relating device, and to a lesser extent fiber, parameters to link parameters.

Conceptually the RF signal could be conveyed over an optical link using any one of the optical carrier's parameters that are analogous to the parameters commonly used with an RF carrier: i.e. the optical carrier's amplitude E_o, frequency v, or phase θ. For specificity, assume an optical plane wave propagating in free space in the z-direction:

$$E(z, t) = E_o \exp\left[j2\pi\left(\frac{zv}{c} - vt + \theta\right)\right]. \tag{1.1}$$

[2] There are techniques, such as wavelength division multiplexing or WDM, by which two or more independent signals can be conveyed over a single fiber. Thus it is possible to use a pair of links, operating at different wavelengths, to provide bi-directional transmission over a single fiber. The individual links, however, are still uni-directional.

[3] There have been attempts to design devices that can both emit and detect light. Initial attempts yielded devices with a considerable compromise in the efficiency of the emitter or detector. However, more recent devices have reduced this combination penalty considerably; see for example Welstand *et al.* (1996).

Means exist in the optical domain that duplicate many of the functions in the RF domain: frequency mixing, LO generation, heterodyne conversion, optical amplifiers – one notable exception is the present lack of any method for hard optical limiting, as there is in the electronic domain. Indeed, optical modulators for each of the three parameters listed above have been demonstrated. However, the technology for optical receivers is at present roughly where RF receivers were at the beginning of the twentieth century.

Virtually all present RF receivers are coherent receivers in which the amplitude or frequency – or in some cases the phase – of the incoming carrier is detected. This is in contrast to the early days of radio when direct detection was the norm – i.e. detection of the presence/absence of the RF carrier without regard to its precise frequency and certainly without any phase information (e.g. Morse code).

Direct detection of an intensity modulated optical carrier is straightforward; as we will see in Chapter 2 all that is required is a photodiode (Yu, 1996). Demodulation of an optical carrier, which has been modulated in either frequency or phase, requires a coherent optical receiver. In turn this requires an optical local oscillator, optical mixer – which can be done in the photodiode – and optical filter. Although coherent optical receivers have been extensively studied (see for example Seeds, 1996; Yamamoto and Kimura, 1981) they have not found widespread application at present, primarily due to the fact that their marginal performance improvement over direct detection does not justify their significant additional complexity.

The results of the coherent optical receiver studies indicated that coherent detection offers greater sensitivity than direct detection. Although coherent detection links require about the same *total* optical power at the photodetector, they require less *modulated* optical power than direct detection for the same signal-to-noise ratio, when used in conjunction with a high optical power local oscillator. This fact was the driving force behind much of the early work on coherent detection. However, more recently, the availability of optical amplifiers, which can be used as optical pre-amplifiers before the photodetector, has permitted direct detection sensitivity to approach that of coherent detection.[4]

Although direct detection is much simpler to implement than coherent detection, it detects only the intensity of the optical wave; all frequency and phase information of the optical carrier is lost. We can see this by examining the intensity of the plane wave example from above. The intensity I (W/m^2) is

$$I = \frac{1}{2}c\varepsilon_0 E_o^2 \qquad (1.2)$$

---

[4] The degree to which the performance of optically pre-amplified direct detection approaches coherent detection depends on many factors; primary among them is the noise figure of the optical pre-amplifier.

or simply the square of the optical wave's amplitude – when the amplitude is real – and where ε_0 is the permittivity and c is the speed of light, both in vacuum. Consequently, amplitude and intensity modulation are not synonymous.[5] One important aspect of this distinction is that the spectrum of the optical waveform for intensity modulation can be much wider than the RF spectrum of the modulating waveform, because the optical spectrum contains harmonics of the modulation waveform generated in the square-law modulation process. This situation is shown diagrammatically in Fig. 1.4(b) by the ellipses that indicate continuation of the sidebands on both sides of the optical carrier. For low modulation indices these harmonics may be negligible, in which case the intensity and amplitude modulated spectra have approximately the same bandwidth. The similarities that intensity and amplitude modulation do share often lead to these terms being used interchangeably in casual discussions. This is unfortunate because the unsupported assumption of equivalence can lead to erroneous conclusions.

Thus intensity modulation of the optical carrier followed by direct detection – which is often abbreviated IMDD – is the universal choice in applications today and will be the focus of this book.

There are two broad categories of optical intensity modulation (Cox *et al.*, 1997). In the simple link example given above, and as shown in Fig. 1.4(a), with *direct modulation*, the electrical modulating signal is applied directly to the laser to change its output optical intensity. This implies that the modulating signal must be within the modulation bandwidth of the laser. As we will see, only semiconductor diode lasers have sufficient bandwidth to be of practical interest for direct modulation. The alternative to direct modulation is *external*, or *indirect*, *modulation*; see Fig. 1.4(c). With external modulation, the laser operates at a constant optical power (i.e. CW) and intensity modulation is impressed via a device that is typically external to the laser. Since there is no modulation requirement on the laser for external modulation, this removes a major restriction on the choice of lasers that can be used for external modulation. Both methods achieve the same end result – an intensity modulated optical carrier – and consequently both use the same detection method, a simple photodetector. As we will discuss in the chapters to follow, there are a number of fundamental and implementation issues concerning these two approaches that give each distinct advantages.

Ideally the electro-optic and opto-electronic conversions at the modulation and photodetection devices, respectively, would be highly efficient, strictly linear and

---

[5] As an example of true optical AM, one could apply the modulation to a Mach–Zehnder modulator biased at cutoff, which will produce double-sideband, suppressed carrier (DSSC) AM of the optical wave. Such a signal can be demodulated by coherently re-injecting the carrier at the receiving end, either optically or by first heterodyning to lower frequencies.

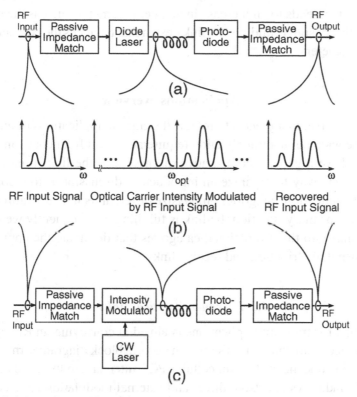

Figure 1.4 Intensity modulation, direct detection links (b) using (a) direct and (c) external modulation.

introduce no noise. Further the devices would maintain these characteristics over all frequencies and for any RF power no matter how large or small.

From the simple link example presented above, we saw that some practical electro-optic devices fall well short of the ideal conversion efficiency goal. As we will see in later chapters, practical devices often also fall short of the other ideal characteristics as well. For example, without proper design practices, we commonly find unacceptable levels of distortion in analog links – which means that one or more of the conversion processes is not strictly linear. Casual link design can also lead to additional noise being added by the optical link, which can reduce the link's ability to convey low level signals. At the other end of the RF power range, practical devices are also limited in the maximum RF power they can handle; typically above a certain RF power, there is no further increase in modulation and the device is said to have saturated.

As we go through this book we will develop the basis for the present limitations of practical devices, then present techniques for reducing these limitations. Our task as a link designer is complicated by the fact that reducing one parameter, such as the noise, often leads to an increase in another parameter, such as the distortion. The "art" of link design is finding *one* link design that balances the competing effect of several parameters.

1.2 Applications overview

There are, of course, many ways to review the current applications of analog optical links. The wide range of application requirements and frequency ranges makes it difficult to have a general comparative discussion of the links used in all such applications. One way to organize an introductory discussion is to recognize that the different application categories emphasize different technical requirements depending on the primary function the link is fulfilling. Consequently we can group fiber optic links into three functional categories that dominate the applications at present: transmit, distribution and receive links.

1.2.1 Transmit optical links

An optical link for transmit applications is aimed at conveying an RF signal from the signal source to an antenna, as shown in generic block diagram form in Fig. 1.5. Applications include the up-link for cellular/PCS antenna remoting and the transmit function of a radar system. Both direct and external modulation have been investigated for radar transmit applications whereas only direct modulation is presently used for cellular/PCS transmit applications.

Since high level signals are involved in transmit, noise is not usually a driving requirement. In radar applications, generally the link needs to convey only a single frequency at a time; consequently distortion is also not a driving requirement. However, in multi-function antennas and cellular/PCS up-links, multiple signals are present simultaneously, so there is the need to meet a distortion requirement, albeit a relatively modest one in comparison to receive applications.

Virtually all transmit applications do require a relatively high level RF signal to drive the antenna. As suggested by Fig. 1.3, most fiber optic links have significant RF-to-RF loss, and it turns out that this high loss also occurs at low frequencies such as UHF. In addition, the maximum RF power at the photodetector end of the link is typically limited by thermal and linearity constraints to about −10 dBm.[6] Consequently for a transmit antenna to radiate 1 W means that 40 dB of RF gain is needed between the link output and the antenna. Further, if this link has a gain of −30 dB, then 20 dBm of input power is necesssary to produce −10 dBm at the link output. However, 20 dBm is above the saturation power of many modulation

---

[6] The unit dBm is power relative to 1 mW, thus −10 dBm represents a power of 0.1 mW.

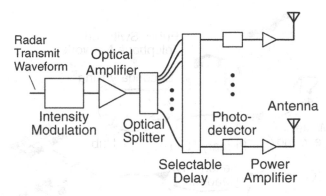

Figure 1.5 Example block diagram of a fiber optic link used for true-time-delay beam steering in a phased array antenna.

devices, which means that a lower drive to the link and consequently a higher gain power amplifier after the link photodetector is typically required. Therefore there is a real need to decrease the RF/optical conversion loss and increase the output power capability of links for transmit applications.

The center frequency of the signal sent to the antenna can be anywhere from 10 MHz to 100 GHz. Complete links have been demonstrated up to 20 GHz. Consequently at center frequencies in this range, the transmit link is typically designed to convey the center frequency without any frequency translation.

The components necessary for higher frequency links have been demonstrated: broadband modulation of a diode laser up to 33 GHz (Ralston *et al.*, 1994), of an external modulator up 70 GHz (Noguchi *et al.*, 1994) and of a photodetector up to 500 GHz (Chou and Liu, 1992) has been reported. However, the efficiencies of these components are such that if they were combined into a link, the link gain without any amplifiers would be rather low. For instance, if the 70 GHz modulator were used in a link with the 500 GHz photodetector with 1 mW of incident optical power, the gain would be approximately -60 dB at 70 GHz. This level of performance only serves to extend the needs mentioned above to include reduced loss at high frequency as well.

1.2.2 Distribution optical links

This type of link is intended to distribute the same RF signal to a multiplicity of sites, such as distributing the phase reference within a phased array radar. The first large scale commercial application of analog fiber optic links was the distribution of cable television (CATV) signals (see for example Darcie and Bodeep, 1990; Olshansky *et al.*, 1989). As shown in Fig. 1.6, the low loss of optical fibers permitted reducing or even eliminating the myriad repeater amplifiers that had been required with coaxial distribution. Like the transmit links, distribution links convey relatively high level

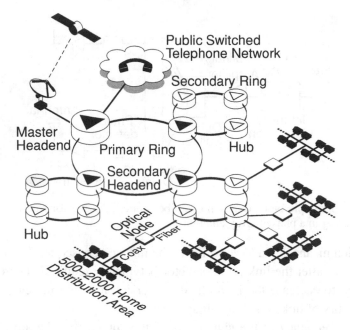

Figure 1.6 Conceptual block diagram of optical-fiber-based CATV distribution system.

signals, consequently link noise is not a driving parameter. Also like some transmit links, distribution links that broadcast the phase reference, such as in a radar, have only a single frequency present at any one time, therefore distortion is not a driving parameter. However, in other distribution applications, such as CATV, multiple carriers – which can be as many as 80 in current CATV systems – are present simultaneously, so distortion becomes a key link parameter. Further, for CATV distribution, the bandwidth is sufficiently wide that both narrow-band and wide-band distortion – two terms that will be defined in Chapter 6 – must be taken into consideration in designing links for this application. Typically, external modulation links are used in the primary ring and direct modulation is used in the secondary ring.

Distribution links by their very nature have a high optical loss that is dominated by the splitting loss in the distribution network, which arises from dividing the optical signal among multiple photodetectors. For example, a fiber network that needs to distribute a signal to 100 locations would have an optical splitting loss of 20 dB, assuming ideal optical splitters that introduce no excess loss in the splitting process. As we will see in Chapter 3, 20 dB of optical loss translates to 40 dB of RF loss between the RF input and any one of the RF outputs. Although the total modulated optical power required is high, the power on each individual photodetector is low.

One convenient way to overcome the high splitting loss is by the use of optical amplifiers. The two basic types of optical amplifiers are semiconductor (see for example O'Mahony, 1988), which are available at either of the principal fiber wavelengths to be discussed in Section 1.3, and solid state[7] (see for example Desurvire, 1994), which although available at either wavelength are at present only commercially viable at the longer wavelength fiber band of 1.55 μm in the form of erbium-doped fibers. Either type of amplifier is capable of 30 to 40 dB of optical gain, but since these amplifiers have equal gain in either direction, optical reflections in links using them must be minimized to avoid spurious lasing.

The most logical location of optical amplification to compensate for splitting loss is to insert a single optical amplifier before the optical splitter. Unfortunately the low saturation power of present optical amplifiers generally rules out this location. Therefore optical amplifiers are more commonly located after the splitter, which means that one amplifier per splitter output is needed instead of one per splitter input. In some distribution applications one set of optical amplifiers is all that is necessary. In other applications with several levels of splitting, one set of optical amplifiers per level of splitting may be needed. Each optical amplifier emits broad bandwidth noise in addition to the amplified coherent (narrowband) light. If this noise is not reduced through filtering, it is possible for subsequent stages of optical amplification to amplify and eventually be saturated by this broadband light.

1.2.3 Receive optical links

These links are designed to convey an RF signal detected by an antenna to an RF receiver located remotely from the antenna. Examples of receive links include the down-link in a cellular/PCS system and the receive mode in a radar (see for example Bowers *et al.*, 1987). A typical block diagram of a directly modulated receive link in a cellular/PCS application is shown in Fig. 1.7.

Since these links are designed primarily for conveying low level signals from the antenna, low noise is one of the primary technical goals for receive links. The low noise of present low frequency, e.g. 100 MHz, links, combined with the high sky noise[8] at low frequencies permits the design and implementation of externally modulated fiber optic links with sufficiently low noise that they can be connected directly to the antenna without the need for any electronic pre-amplification.

As the required link operating frequency increases, the sky noise decreases and the link noise increases, primarily due to the higher link loss. Consequently virtually all high frequency receive links require a low noise RF pre-amplifier between the antenna and the link input. In principle the high link noise can be reduced to nearly

---

[7] It is important to keep in mind that unlike the electronics field, the laser field uses semiconductor and solid state to refer to two distinctly different types of lasers. Semiconductor lasers are almost exclusively excited or pumped electrically by passing an electrical current through a diode junction whereas solid state lasers are universally pumped optically.

[8] Sky noise is used to refer collectively to noise sources that are external to the antenna. The dominant components of sky noise are atmospheric noise, which is distributed primarily throughout the RF spectrum below 50 MHz, and cosmic noise, which limits a receiver's minimum detectable signal over the frequency range 10 to 300 MHz (Uitjens and Kater, 1977).

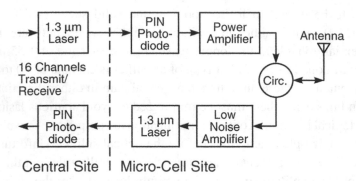

Figure 1.7 Representative block diagram showing remoting of a wireless communication antenna via optical fiber.

the electronic pre-amplifier noise if the gain of this pre-amplifier is sufficiently high. However, this approach is limited in practice by the need to meet simultaneously other link RF requirements, such as a distortion requirement. Thus one of the key needs for receive applications is to develop links with sufficiently low noise. As we shall see in Chapter 5, achieving low link noise is fundamentally tied to reducing the link RF loss.

In addition to noise, another important receive link parameter is distortion. One important aspect of many receive links that does simplify the distortion problem is that broadband antennas are rare, and those that are broadband achieve a wide bandwidth at a severe tradeoff in sensitivity. Consequently most receive links need only an octave (2:1) bandwidth or less, which means that wide-band distortion can be filtered out and that narrow-band distortion is the dominant factor. This is an important consideration when one is faced with an application that requires lower distortion than is available from the intrinsic electro-optic device. In such cases, as we will see in Chapter 6, linearization can be used to reduce the distortion. However, all the known broadband linearization techniques – i.e. those that reduce both the wide-band and narrow-band distortion – invariably increase the link noise. In contrast, narrow-band-only linearization techniques, which invariably increase the wide-band distortion, do not suffer a significant noise penalty (Betts, 1994). There is a need for wide-band receive links, and they present one of the principal unsolved challenges to link design: achieving low narrow- and wide-band distortion while simultaneously achieving low noise.

1.3 Optical fibers

The choice of wavelength for the optical carrier depends on the availability of the electro-optic devices and fiber with the required performance.

Many of the electro-optic devices are fabricated in semiconductors, which permits great flexibility in the choice of operating wavelength. Consequently photodetectors are presently commercially available for any wavelength from the near ultra-violet to well into the infra-red. The wavelength range for diode lasers is not quite as broad but presently diode lasers are available from the blue into the near infra-red. External modulators fabricated in the most common electro-optic material, lithium niobate, are transparent over about the same spectral range as diode lasers. However, photorefractive effects limit the maximum usable optical power in lithium niobate to wavelengths longer than about 1 μm. Solid state lasers, which are commonly used as the CW source for external modulation, are generally available only at specific wavelengths, such as 1.06 or 1.319 μm for neodymium YAG lasers, although titanium sapphire lasers are tunable over the wavelength range 0.7 to 1 μm. Consequently the electro-optic devices only broadly constrain the wavelength choice.

One of the fiber's key wavelength parameters is optical loss. Figure 1.8 is a plot of the loss vs. wavelength for a typical, silica-based optical fiber. The visible portion of the spectrum is between about 0.4 and 0.7 μm, which extends off the left end of the scale in Fig. 1.8. Consequently all the wavelengths used for optical communications at present are in the near infra-red portion of the spectrum. As this plot makes clear, Rayleigh scattering from the atoms constituting the glass itself sets the lower limit on the fiber's optical loss. The specific value of this limit depends on the optical design of the fiber and will be discussed in more detail below. The figure also shows several wavelength bands where the loss increases due to the residual effects of impurity absorption – most importantly the OH radical – in the fiber. More recent fibers have virtually eliminated the dominant OH absorption peak between 1.3 and 1.55 μm.

The result of combining the device and fiber constraints is that there are three primary wavelength bands that are used in fiber optic links. By far the dominant wavelengths in use at present are located in bands around 1.3 and 1.55 μm. From Fig. 1.8 it would appear that the best wavelength band would be the one around 1.55 μm. Indeed, this is where the lowest loss is and hence this would be the best wavelength to use for long length links. However, the band around 1.3 μm has also been used extensively because originally this was the only band where the chromatic dispersion – i.e. the change in propagation velocity with wavelength – was zero, which is important for high frequency and/or long length links. The *chromatic* dispersion at 1.55 μm is typically 17 ps/(nm km) unless special processing steps are taken to fabricate 1.55 μm fiber with zero wavelength dispersion, the so-called dispersion shifted fiber, which also has low attenuation. A further attraction of the 1.55 μm band is the availability of fiber optic amplifiers, which so far have not proven feasible in the other wavelength bands. Thus by operating at

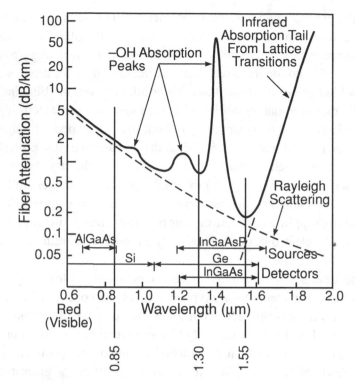

Figure 1.8 Loss vs. wavelength for silica optical fiber and the wavelength ranges of some common electro-optic device materials.

either one of these longer wavelengths, the fiber offers a nearly ideal transmission medium.

The first fiber optic links operated around 0.85 μm, this wavelength being chosen primarily because it corresponded to the availability of the first laser diodes. However, as we have just discussed, this wavelength offers neither of the fiber advantages in terms of optical loss and bandwidth of the longer wavelengths. Still 0.85 μm is of interest because it offers the nearest term prospect for integration of gallium arsenide, GaAs, electronics with diode lasers, which at 0.85 μm are made in the GaAs material system.

An optical fiber conveys light by confining it to a core region, which has a slightly higher refractive index than the surrounding cladding thereby confining the light to the core via total internal reflection. Figure 1.9 presents photographs of the ends of three types of typical optical fibers where the core has been illuminated to make it stand out more clearly. The left photograph shows a *single mode* fiber in which the core is so small, typically ~5 to 8 μm in diameter, that the only possible path

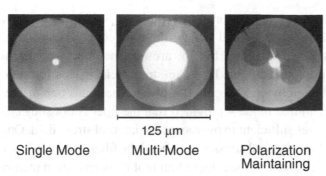

125 μm

Single Mode   Multi-Mode   Polarization
                                        Maintaining

Figure 1.9 End view of a single mode, multi-mode and polarization maintaining optical fiber with the core illuminated. The outer diameter of the cladding is 125 μm in all cases.

is the one in which light propagates straight down the core fiber, i.e. without any reflections off the core–cladding interface.

An alternative is the *multi-mode* fiber shown in the center photograph of Fig. 1.9 where the larger core, typically 50 to 62 μm in diameter, permits multiple light paths to propagate down the fiber: the straight path plus multiple paths that involve one or more reflections off the core–cladding interface. Multi-mode fiber typically supports thousands of optical modes. The larger core makes for more efficient fiber to device coupling than is possible with single mode fiber. However, the multiple modes travel different paths and consequently have different propagation times through the fiber. This leads to *modal* dispersion, i.e. different propagation velocities for the different modes. Typical modal dispersion for multi-mode fibers at 0.85 μm is 90 ps/(nm km), which is more than five times the chromatic dispersion of single mode fiber at 1.55 μm. Consequently for all high performance applications, single mode fiber is the choice and it will be assumed throughout this book, unless noted otherwise.

Although the light from lasers is linearly polarized, standard single mode fibers do not maintain the polarization of the guided light. In some applications, most notably external modulators, a fixed polarization of light at the modulator input is required to maintain maximum modulation efficiency. In principle, if left undisturbed, standard single mode fibers would maintain the state of polarization of the light launched into them. However, this feature is not usable in practice because even small movements of the fiber, or microbends, will change the stresses within the fiber, thereby altering the polarization state of the propagating light. Therefore, to meet the needs of external modulators that are placed remotely from the CW laser, special single mode fibers have been developed that will maintain the polarization state of the propagating light.

The most common method for implementing a *polarization maintaining* fiber[9] is to induce a known stress field around the fiber core. Glass rods with a slightly different thermal expansion coefficient are one method of doing this, as shown in the right photograph of Fig. 1.9. The stress field makes the two polarization modes distinct, thereby ensuring that light launched into one of the polarization modes will stay in that mode – provided that the fiber is not subjected to external mechanical stresses sufficient to overcome the internal stress field. One easy way to impose external mechanical stresses is to bend the fiber. Typically, as a polarization maintaining fiber is bent, we see degradation of its polarization maintaining ability before we see significant degradation of its optical loss.

As was pointed out in Fig. 1.3, the zero length link loss, or conversion loss, is one of the main design issues for fiber links. One contribution to the conversion loss, which we can deal with here, is the coupling loss between the optical fiber and the electro-optic device. This problem arises through the combined effects of the small fiber core and the relatively weak confinement of the light in the core. Both of these factors are related to minimizing the optical fiber loss.

To increase the fiber to electro-optic coupling, we want to make the core as large as possible. A large core would also reduce the loss due to scattering at the core–cladding interface. The way to satisfy both these design objectives, which is in universal use at present, is to use a small core–cladding index step, thereby decreasing the confinement and spreading the light out. This indeed does decrease the requirements on core–cladding scattering loss and increases the optical mode size.

However, the tradeoff is that the core optical mode has a significant evanescent tail that propagates in the cladding. Consequently bending the fiber would require the light on the outside of the bend to exceed the speed of light in the cladding. Since this is not possible, some of the light radiates away from the core at a bend and this represents loss. One measure of the sensitivity to microbends is the loss vs. fiber bend radius. Measurements of loss for one turn around a mandrel vs. the mandrel radius for three types of typical fibers are shown in Fig. 1.10.

The result of the above considerations is a classic design tradeoff: *decrease* the index step to get lower scattering loss and a larger core, versus *increase* the index step to get higher immunity to microbending losses. For example at 1.3 μm, a common combination of design and performance parameters is a core-to-cladding index step of 0.36% that results in core diameters of about 8 μm, optical loss of 0.35 dB/km and modest microbending sensitivity. So-called confined core fibers are available which have an index step of 1.5% and a smaller core diameter – on the

---

[9] This should not be confused with a *polarizing* fiber, which would take an unpolarized input and – after a sufficient length – produce a polarized output.

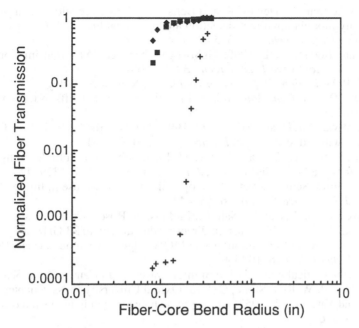

Figure 1.10 Optical loss vs. bend radius of three types of common single mode optical fibers. ♦ Polarization maintaining, ■ confined core, + standard.

orders of 5 μm – that results in much lower microbending sensitivity, but higher propagation loss – typically 0.55 dB/km.

The fiber-to-device coupling efficiency is determined not only by the core diameter, but also by the diameter, shape and degree of collimation of light to or from the electro-optic device. The smaller the core–cladding index step, the higher the collimation that is required for the fiber to capture the light into the core. For typical single mode fibers, the acceptance angle is about 7 degrees and for confined core fiber it is typically 12 degrees. As we will see, these relatively low values of acceptance angle are not well matched to diode lasers, whose output beam is typically elliptical with divergence angles of 10 and 30 degrees.

References

Betts, G. 1994. Linearized modulator for suboctave-bandpass optical analog links, *IEEE Trans. Microwave Theory Tech.*, **42**, 2642–9.

Bowers, J. E., Chipaloski, A. C., Boodaghians, S. and Carlin, J. W. 1987. Long distance fiber-optic transmission of C-band microwave signals to and from a satellite antenna, *J. Lightwave Technol.*, **5**, 1733–41.

Chou, S. and Liu, M. 1992. Nanoscale tera-hertz metal-semiconductor-metal photodetectors, *IEEE J. Quantum Electron.*, **28**, 2358–68.

Cox, C., III, Ackerman, E., Helkey, R. and Betts, G. E. 1997. Techniques and performance of intensity-modulation direct-detection analog optical links, *IEEE Trans. Microwave Theory Tech.*, **45**, 1375–83.

Darcie, T. E. and Bodeep, G. E. 1990. Lightwave subcarrier CATV transmission systems, *IEEE Trans. Microwave Theory Tech.*, **38**, 524–33.

Desurvire, E. 1994. *Erbium-Doped Fiber Amplifiers*, New York: Wiley.

Gowar, J. 1983. *Optical Communication Systems I*, Englewood Cliffs, NJ: Prentice Hall, Section 16.2.1.

Noguchi, K., Miyazawa, H. and Mitomi, O. 1994. 75 GHz broadband Ti:LiNbO$_3$ optical modulator with ridge structure, *Electron. Lett.*, **30**, 949–51.

Olshansky, R., Lanzisera, V. A. and Hill, P. M. 1989. Subcarrier multiplexed lightwave systems for broad band distribution, *J. Lightwave Technol.*, **7**, 1329–42.

O'Mahony, M. 1988. Semiconductor laser optical amplifiers for use in future fiber systems, *J. Lightwave Technol.*, **6**, 531–44.

Ralston, J., Weisser, S., Eisele, K., Sah, R., Larkins, E., Rosenzweig, J., Fleissner, J. and Bender, K. 1994. Low-bias-current direct modulation up to 33 GHz in InGaAs/GaAs/AlGaAs pseudomorphic MQW ridge-waveguide devices, *IEEE Photon. Technol. Lett.*, **6**, 1076–9.

Seeds, A. J. 1996. Optical transmission of microwaves. In *Review of Radio Science 1993–1996*, ed. W. Ross Stone, Oxford: Oxford University Press, Chapter 14.

Taylor, J. H. and Yates, H. W. 1957. Atmospheric transmission in the infrared, *J. Opt. Soc. Am.*, **47**, 223–6.

Uitjens, A. G. W. and Kater, H. E. 1977. Receivers. In *Electronics Designers' Handbook*, 2nd edition, ed. L. J. Giacoletto, New York: McGraw-Hill Book Company, Section 23.

Welstand, R. B., Pappert, S. A., Sun, C. K., Zhu, J. T., Liu, Y. Z. and Yu, P. K. L. 1996. Dual-function electroabsorption waveguide modulator/detector for optoelectronic transceiver applications, *IEEE Photon. Technol. Lett.*, **8**, 1540–2.

Yamamoto, Y. and Kimura, T. 1981. Coherent optical fiber transmission systems, *IEEE J. Quantum Electron.*, **17**, 919–34.

Yu, P. K. L. 1996. Optical receivers. In *The Electronics Handbook*, ed. J. C. Whitaker, Boca Raton, FL: CRC Press, Chapter 58.

2

Link components and their small-signal electro-optic models

2.1 Introduction

In this chapter we develop the small-signal relationships between the RF and optical parameters for the most common electro-optic devices used in intensity modulation, direct detection links. There are numerous device parameters we could use for this task; we concentrate here – as we will throughout this book – on those parameters that can be measured and selected by the *link* designer – as opposed to those parameters that can only be measured and controlled by the *device* designer.

To provide the basis for comparing these and future devices, we develop a figure of merit for optical modulators and detectors: the RF-to-optical incremental modulation efficiency for modulation devices and its converse the optical-to-RF incremental detection efficiency for photodetection devices. These efficiencies are useful in link design because they provide a single parameter for evaluating device performance in a link that represents the combined effects of a device's optical and electrical parameters. Further, by using the same parameter for both direct and external modulation devices, we begin the process – which will carry on through much of the book – of using a single set of tools for evaluating both types of links.

The most common electro-optic devices in use for links today are the in-plane diode laser, both Fabry–Perot and DFB, for direct modulation, the Mach–Zehnder modulator for external modulation and a photodiode for photodetection. Thus on a first reading, one may want to focus on these devices. However, other direct and external modulation devices are included not only because they may become more important in the future, but also because they give us a chance to demonstrate the versatility, and reinforce the technique, of the analytical approach being presented here.

2.1.1 Notation

In the developments to follow, it is important to establish a subscript convention for distinguishing among a number of forms for each of the variables. The subscript of a parameter makes a general variable specific to a device; for example the general variable current is denoted by i, whereas the current specific to the laser is i_L. There are four forms of each variable that we need in our analysis. We distinguish among them using the convention adopted by the IEEE (IEEE, 1964). One variable form is the dc or bias point component, which is represented by an upper-case symbol and an upper-case subscript. Continuing the example from above, the laser dc bias current is denoted by I_L. The modulation component is denoted by lower-case symbols and lower-case subscripts; e.g. i_ℓ. The total value of the variable, at any instant in time, is simply the sum of the dc and modulation components; the total instantaneous value is denoted by a lower-case symbol and an upper case subscript, e.g. i_L. Continuing with the laser current example, the relationship among these variables can be written in equation form as

$$i_L = I_L + i_\ell. \qquad (2.1)$$

Of course in general all the preceding are functions of time, so strictly speaking we should show the time dependence. However, we drop explicit portrayal of the time dependence unless it is required for the discussion. The final variable form we need to distinguish is the log to the base 10 of the modulation component. For this we deviate from the IEEE standard referenced above and use upper-case symbols with lower-case subscripts, i.e. $I_\ell = 10 \log_{10} i_\ell$. The units of I_ℓ are dB (decibels)[1]; the normalizing factor, e.g. mA, optical mW, etc., will be given when each usage is first presented.

2.2 Modulation devices

2.2.1 Direct modulation

With direct modulation, the modulation signal directly changes the intensity of the laser output. Conceptually we could choose any laser for direct modulation; the choice is primarily influenced by the bandwidth, optical wavelength and efficiency of impressing the modulation onto the laser's output. The only laser to date that has demonstrated sufficient bandwidth and efficiency to be of practical interest for direct modulation is the semiconductor diode laser. These lasers come in many configurations: for example with the cavity in the plane of the semiconductor wafer or perpendicular to it, with planar end facets or wavelength-selective gratings to

---

[1] Throughout this book we will be working in both the optical and electrical domains. While the dB is valid in both these domains, it is not valid across these two domains, for reasons that will become clear in later chapters. For example, we will show that 1 dB of optical loss results in 2 dB of RF link loss.

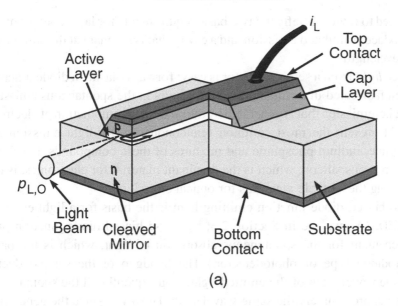

(a)

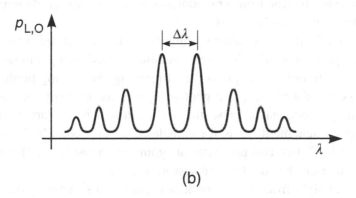

(b)

Figure 2.1 (a) Diagram showing the principal components of a Fabry–Perot diode laser whose laser cavity is in the plane of the semiconductor wafer. (b) Idealized spectral output of a Fabry–Perot diode laser; spectra from actual Fabry–Perot lasers are generally not symmetrical or equally spaced. (Cox, 1996, Fig. 57.1. © 1996 CRC Press, reprinted with permission.)

provide the optical reflections, etc. Since our goal here is to develop the basic device models upon which we can build the link design parameters in future chapters, we choose a simple laser design for the emphasis of the discussion. We then introduce variations to this laser design – and the link performance refinements they permit.

2.2.1.1 Fabry–Perot diode laser

A diagram of a generic in-plane, edge-emitting Fabry–Perot diode laser is shown in Fig. 2.1(a). The principal parts are a p-n diode junction and an optical waveguide with partially reflecting mirrors at either end. To appreciate how this structure makes a laser, we need to review briefly the two basic requirements for laser operation: the ability to produce stimulated emission and a cavity that is resonant at the stimulated emission wavelength.

To produce *light* from a semiconductor is easy: forward biasing a diode junction that has been fabricated in a suitable material results in the spontaneous emission of radiation, i.e. emission that is generated by the random recombination of electrons and holes. At present the most common semiconductors for light emission are gallium arsenide, indium phosphide and mixtures of these compounds. For fairly fundamental reasons silicon, which is the dominant material for electronics, is not currently among the suitable materials for optical emission.

A forward biased diode junction emitting light is the basis for a light emitting diode, or LED. (As we see in Section 2.3, a reverse biased diode junction provides a mechanism for the spontaneous absorption of light, which is the basis of the photodiode type of photodetector.) The bandgap of the semiconductor determines the wavelength of the emitted light. Consequently all the spontaneous photons have approximately the same wavelength. However, since the generation of a spontaneous photon came from a random process there is no correlation among the phases of the spontaneous photons.

In a laser we have *l*ight *a*mplification by *s*timulated *e*mission of *r*adiation. Unlike the spontaneous processes just discussed, stimulated processes – emission and absorption – result in *identical* copies of the triggering event being produced. In stimulated emission of a photon, one photon triggers the emission of a second, *identical* photon. Thus one way to view stimulated emission is as a form of positive gain. An analogous statement holds for stimulated absorption, which results in negative gain or loss. These two processes are going on in parallel so the one that dominates is the one that has the larger reservoir of carriers.

To produce *laser* light from a semiconductor requires a structure where an inverted population consisting of both electrons and holes can be set up so that stimulated emissions dominate over stimulated absorptions. In such a structure, a few spontaneous emissions (LED) can trigger stimulated emissions (laser), each of which in turn can trigger additional stimulated emissions and so forth. In a semiconductor, the inverted population can be created and maintained by passing a dc current through the diode junction that has provisions to confine the electrons and holes. Thus the diode junction provides one of the two basic requirements for laser operation.

The second criterion for laser operation is that these stimulated emissions must occur within an optical cavity that is resonant at the stimulated emission wavelength. The most common type of optically resonant cavity for diode lasers is one formed by two plane, parallel mirrors. Such a cavity is referred to as a Fabry–Perot after the original developers of this type of optical interferometer. In a Fabry–Perot cavity, with a vacuum between the mirrors, resonances exist for any wavelength, λ_o, which after reflections off of both of the mirrors are again in phase with the original wave. For a cavity formed by two mirrors spaced a distance l apart, the condition for the free-space resonant wavelengths of a Fabry–Perot cavity can be expressed as

$$2l = m\lambda_o. \tag{2.2}$$

In semiconductor lasers the cavity length, l, is typically at least 100 times the stimulated emission wavelength, so that m is a large integer. Therefore the wavelength spacing between two adjacent longitudinal modes (m and $m + 1$) within the cavity, $\Delta\lambda$, is approximated by

$$\Delta\lambda \approx \frac{\lambda^2}{2l}. \tag{2.3}$$

When the cavity is formed in a semiconductor with index of refraction n, the wavelength in the cavity is related to the free-space wavelength as $\lambda_o = n\lambda$. However, when we refer to a "1.3 μm" laser, we are referring to its free-space wavelength. Thus we need to convert the Fabry–Perot resonances within the semiconductor, (2.3), into the corresponding free-space wavelengths. To do so we simply substitute $\lambda = \lambda_o/n$ for both wavelength terms in (2.3); the result is

$$\Delta\lambda_o \approx \frac{\lambda_o^2}{2nl}. \tag{2.4}$$

In an in-plane Fabry–Perot diode laser operating at 1.3 μm, a common cavity length is 300 μm and the refractive index is about 3.1, so the longitudinal mode spacing would be 0.91 nm. This is an incredibly large wavelength spacing compared to other types of lasers where the cavity is centimeters to meters long and the corresponding intermodal spacing is less than 10 picometers. The short cavity length is possible because the optical gain of semiconductors is so high relative to lasers using other materials. As discussed in Chapter 4, such a short cavity is one of the reasons for the higher modulation frequency response of diode lasers. However, even with this relatively short cavity and wide mode spacing, the optical gain spectrum of semiconductors is wide enough to permit lasing on several modes simultaneously, as shown by the spectral plot of Fig. 2.1(b) for a typical Fabry–Perot laser.[2] There are techniques for limiting the lasing to a single longitudinal mode, and this is one of the laser variations that is discussed below. For further information on the various aspects of lasers in general and diode lasers in particular, there are numerous books to which the reader can refer; among them are Thompson (1980), Agrawal and Dutta (1986) and Coldren and Corzine (1995).

---

[2] Although Fig. 2.1(b) shows a modal pattern that is symmetric vs. wavelength, in practice the asymmetry with respect to wavelength of the semiconductor gain curve results in an asymmetric distribution of optical power into the various modes.

To get some of the light out of an in-plane laser cavity at least one, and often both, of the mirrors is only partially reflecting. For an in-plane laser, it turns out to be easy to fabricate the required partially reflecting mirrors in a single crystal material such as a semiconductor since these materials crack or cleave along crystal planes, thereby ensuring automatically that the resulting surfaces are optically smooth and parallel to each other. A convenient amount of reflection is also easy to obtain: the Fresnel reflection that results from the differences in dielectric constants at the semiconductor-to-air interface is about 65% in terms of intensity. The high optical gain of semiconductor lasers means that they can still operate while losing 1/3 of their optical power on each pass through the cavity.

To ensure a linear optical power vs. current function and a stable spatial optical mode, the light must be confined horizontally and vertically in a single-spatial-mode waveguide between the mirrors. Early diode lasers relied on gain guiding to confine the light: the optical gain region had a sufficiently higher index of refraction to confine the light. Although gain guiding is simpler from a fabrication standpoint, this needed to be traded off against performance compromises. Present in-plane diode lasers for direct modulation are index guided, which means that the light is confined by the geometry and composition of the semiconductor layers. Typical dimensions for the active layer are 1 to 2 μm wide and 0.2 to 0.5 μm thick. The fact that these two dimensions are not equal results in the in-plane laser beam having an elliptical cross section, rather than a circular one. Such an elliptical emission pattern is one of the main reasons for inefficiency in coupling a diode laser output to the circular mode of optical fibers. Despite these inefficiencies, the typical fiber-coupled optical power can range from 10 to 100 mW.

From a link design viewpoint, one of the useful graphs of laser performance is of the laser's optical power, P, vs. laser current, I. A representative P vs. I curve for a diode laser is shown in Fig. 2.2. It is possible to derive this curve and the important features of it such as the threshold current and slope efficiency – which will be discussed below – from a coupled pair of differential equations that are known in the laser device field as the rate equations. However, this derivation does not add much insight for the link designer. Hence we choose to develop here the P vs. I curve based on plausibility arguments and defer the derivation of the dc rate equations to Appendix 2.1.

As the current is increased from zero, initially the optical output is dominated by spontaneous emission and the output increases slowly with current. At low currents the probability of stimulated emission is low and so most of the photons are absorbed by the semiconductor. An alternative way to view this situation is that the positive optical gain – as represented by the stimulated emission – is less than the negative optical gain (optical losses) – represented by the photons lost through absorption and emission. Increasing the current increases the number of photons

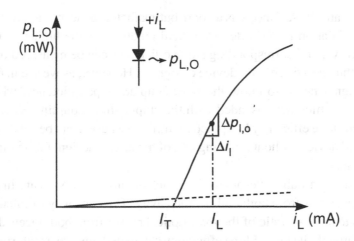

Figure 2.2 Representative plot of a diode laser's optical power, $p_{L,O}$ vs. the current through the laser, i_L, with the threshold current, I_T, and a typical bias current, I_L, for analog modulation.

generated via both the spontaneous and stimulated processes. Eventually a *threshold* current is reached where the number of photons generated equals the number of photons absorbed. Consequently, at threshold the optical gain compensates for all the optical losses. At threshold, the typical optical output of an in-plane laser is 10 to 100 μW. Above the threshold current level, the optical gain is greater than the optical losses. As a result stimulated emission begins to dominate the optical output and the output increases at least 100 times more rapidly with laser current than it did below threshold.

The threshold current, I_T, can be determined from the laser P vs. I curve by extrapolating the straight line portion of the stimulated-emission dominated region down to where it intersects the current axis. This current is an important measure of laser performance because if it is too high, the heating caused by passing a large current through the junction precludes operating the laser CW. Typical values of threshold current for present commercially available in-plane lasers are in the 5 to 50 mA range depending on laser active area; however, threshold currents significantly lower, around 0.1 mA, have been reported for experimental devices (Yang *et al.*, 1995; Chen *et al.*, 1993). As a practical matter, the threshold current increases with temperature, which usually necessitates some form of temperature control.

The slope efficiency, s_ℓ, is a laser figure of merit that is used extensively in link modeling. This is simply the incremental slope of the P vs. I curve at a given laser bias current, I_L, and is defined as

$$s_\ell(i_L = I_L) = \frac{dp_\ell}{di_\ell}. \tag{2.5}$$

The units of s_ℓ are W/A. Since s_ℓ is a contributing factor to the efficiency with which electrical modulation is converted to optical modulation, the higher s_ℓ, the better. The P vs. I curve, and corresponding slope efficiency, can be measured on the bare laser diode chip, and often is by device designers. However, as we see in Chapter 3, the link designer needs to know the fiber-coupled slope efficiency, which is the bare chip slope efficiency cascaded with the chip-to-fiber coupling efficiency. The fiber-coupled slope efficiency has practical utility because it can be easily measured on a packaged device without knowing any of the laser's design, fabrication or fiber coupling details.

Lasers that use a Fabry–Perot cavity – such as diode lasers – can, in principle, have laser emissions from both ends of the cavity. Thus one way to calculate slope efficiency is to take the ratio of the *total* optical power from both facets divided by the current through the laser. Slope efficiency calculated this way is referred to as the total or double-ended slope efficiency. However, for a link to benefit from double-ended slope efficiency, the outputs from both facets would have to be coupled into the *same* fiber, which is rarely, if ever, done. Consequently for link calculations the single-ended slope efficiency is the relevant parameter; it is the ratio of the optical power from one facet to the current through the laser. If the laser facets have been coated such that the optical power from one facet is greater than the other, then the single-ended slope efficiency must be calculated using the optical power emitted from the facet that is coupled to the link photodetector. In this book we assume that slope efficiency refers to the single-ended slope efficiency, unless otherwise noted.

While the differential expression for the slope efficiency, (2.5), is quite useful, it obscures some facts about the slope efficiency. For example, from the differential form, it is not evident that the slope efficiency has a maximum value or that it has a wavelength dependence. To expose these aspects of the slope efficiency, we need to express it in terms of a closely related parameter, the external differential quantum efficiency, η_ℓ, which is the ratio of the change in the number of emitted photons to the change in the number of injected electrons. The expression for the slope efficiency at a given bias current, I_L, in terms of the quantum efficiency is

$$s_\ell(I_L) = \frac{\eta_\ell h c}{q \lambda_o}, \tag{2.6}$$

where q is the electron charge, λ_o is the free-space wavelength, h is the Planck constant and c is the velocity of light in vacuum.

Equation (2.6) makes clear a couple of points about the slope efficiency; one is that there is an upper bound on slope efficiency. This maximum slope efficiency arises from the fact that we can get at most one photon per electron, i.e. $\eta_\ell \leq 1$ when the laser facets are asymmetrically coated so that all the optical power is emitted from one facet.

The other aspect of slope efficiency that we can see from (2.6) is its wavelength dependence. To appreciate the basis for the slope efficiency wavelength dependence, recall that (2.5) is the ratio of optical power to electrical current. And whereas the current is independent of wavelength, the energy of the photons comprising the optical power has a wavelength dependence. Thus the *same* differential quantum efficiency will lead to two *different* slope efficiencies at two different wavelengths. For example $\eta_\ell = 75\%$ yields $s_\ell = 0.71$ W/A at $\lambda_0 = 1.3$ μm but at $\lambda_0 = 0.85$ μm, $s_\ell = 1.09$ W/A. As this example makes clear, although the upper bound on η_ℓ is one, the upper bound on s_ℓ is a number greater or less than 1 W/A, depending on the wavelength.

To get a feel for both the upper bound and the wavelength dependence, equation (2.6) is plotted in Fig. 2.3 for the case $\eta_\ell = 1$ and over the wavelength range that includes all those in common use today. Also plotted on Fig. 2.3 are the reported fiber-coupled slope efficiencies for several research and commercially available diode lasers. In addition to data on in-plane Fabry–Perot lasers, Fig. 2.3 also includes data on distributed feedback and vertical cavity lasers, which are discussed in Sections 2.2.1.2 and 2.2.1.3, respectively. The corresponding bandwidths for each of the devices in Fig. 2.3 are presented in Chapter 4.

As we can see, the fiber-coupled slope efficiencies obtained in practice are considerably lower than the maximum theoretical value. The primary reason for the low slope efficiency is the low coupling efficiency between the diode laser output and the fiber. In turn, the low coupling efficiency is due to the differences in optical mode size, shape and divergence between the laser waveguide and the fiber core. One of the other diode laser variations to be discussed below offers higher fiber coupling efficiency via its improved beam shape.

For the link models to be developed in Chapter 3, we need a small-signal expression for the relationship between the electrical modulation power applied to the diode laser and the modulated optical output power from the laser, $p_{\ell,o}$. It turns out that the best measure for the amount of electrical modulation is its available power, $p_{s,a}$ which is defined as the power delivered to a load whose impedance *matches* that of the modulation source. If we model the modulation source as a voltage source of magnitude v_s in series with an impedance R_S, then the available power – in Watts – is

$$p_{s,a} = \frac{v_s^2}{4R_S}. \tag{2.7}$$

For clarity of presentation, we limit the present discussion to modulation frequencies that are sufficiently low to permit us to neglect the reactive components of the diode laser impedance. This leaves only the resistive component, which we denote by R_L. Typically R_L for an in-plane laser is less than R_S, so we need some

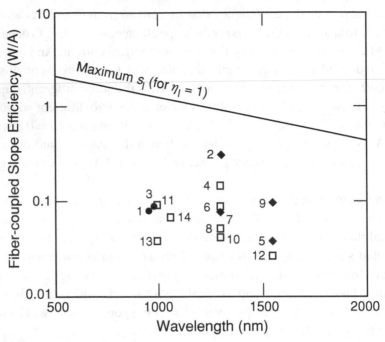

Figure 2.3 Plot of the maximum fiber-coupled slope efficiency vs. wavelength. Also shown are reported slope efficiencies for several research and commercially available diode lasers. ● *VSCEl*, ◻ *FP*, ◆ *DFB*.

Point	Authors and date	Freq. (GHz)
1	M. Peters *et al.*, 1993	2.8
2	Fujitsu, 1996	3
3	B. Moller *et al.*, 1994	10
4	T. Chen *et al.*, 1990	12
5	Y. Nakano *et al.*, 1993	12
6	W. Cheng *et al.*, 1990	13
7	T. Chen *et al.*, 1994	18
8	H. Lipsanen *et al.*, 1992	20
9	T. Chen *et al.*, 1995	20
10	R. Huang *et al.*, 1992	22
11	L. Lester *et al.*, 1991	28
12	Y. Matsui *et al.*, 1997	30
13	J. Ralston *et al.*, 1994	33
14	S. Weisser *et al.*, 1996	40

form of impedance matching between the modulation source and the laser to satisfy the definition of available power. A simple, but as we will see inefficient, way to implement the impedance match is to add a resistor, R_{MATCH}, in series with the laser. The value of this resistor is chosen such that $R_S = R_L + R_{MATCH}$. Putting all the foregoing together results in the circuit shown in Fig. 2.4.

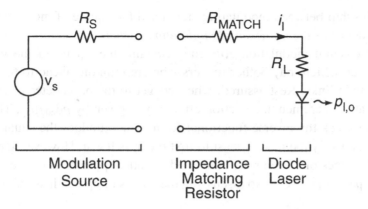

Figure 2.4 Schematic diagram of the lumped-element, low frequency circuit model of a modulation source; an ideal, impedance matching resistor and a diode laser.

Our goal now is to develop an expression for the relationship between $p_{\ell,o}$ and $p_{s,a}$. From (2.5) we already know that we can relate $p_{\ell,o}$ to the laser current i_ℓ, namely,

$$p_{\ell,o} = s_\ell i_\ell, \tag{2.8}$$

where we have dropped the explicit dependence of slope efficiency on bias current. The laser current, i_ℓ, depends on the source voltage minus any voltage drop across the diode laser junction. It can be shown that when the laser is biased above threshold, the incremental voltage drop across the diode laser junction is negligible compared with the incremental voltage drop across R_L (Agrawal and Dutta, 1986, pp. 23–69, 184–186 and 256–263). Using this approximation, a straightforward analysis of the circuit shown in Fig. 2.4 yields the following relationship between i_ℓ and v_s:

$$i_\ell = \frac{v_s}{R_S + R_L + R_{\text{MATCH}}}. \tag{2.9}$$

Substituting the square of (2.9) into the square of (2.8) yields

$$p_{\ell,o}^2 = \frac{s_\ell^2 v_s^2}{(R_S + R_L + R_{\text{MATCH}})^2}. \tag{2.10}$$

We now need to divide (2.10) by the equation for available power, (2.7). This can be done by recalling that under the condition where the laser is matched to the source, $R_S = R_L + R_{\text{MATCH}}$, so the denominator of (2.10) can be written as $4R_S^2$. Using this substitution when dividing (2.10) by (2.7) leaves only R_S in the denominator, which again using the fact that $R_S = R_L + R_{\text{MATCH}}$, lets us write the result as

$$\frac{p_{\ell,o}^2}{p_{s,a}} = \frac{s_\ell^2}{R_L + R_{\text{MATCH}}}. \tag{2.11}$$

This relationship between modulating signal and the square of modulated optical power we define as the *incremental modulation efficiency*.

The incremental modulation efficiency contains the square of the modulated optical power, which may bother readers who are thinking ahead to the question of distortion in links. Rest assured, when we get to the photodetector model, we will see that its incremental detection efficiency is given by p_{load}/p_{od}^2. That is, the photodetector has the inverse functional form, which undoes the squaring at the laser. Consequently the link, at least to first order, is linear. However, the fact that electrical voltages or currents are converted to optical powers, and vice versa, will have an impact when we get to discussing the effects of optical loss on the RF gain of the link.

The alert reader may also question why we have R_{MATCH} in (2.11) given our need to make the electrical-to-optical conversion efficiency as high as possible. Would not leaving it out be a simple way to increase this efficiency? The answer is yes it would and in fact this is sometimes done. However, we defer further discussion of this point to the end of Section 3.2.1.

2.2.1.2 Distributed feedback diode laser

As was mentioned above, the optical spectrum of a Fabry–Perot diode laser is composed of several wavelengths, i.e. all those wavelengths that satisfy (2.2) and fall within the semiconductor gain spectrum. This can present a problem in applications where the fiber length or the modulation bandwidth are such that wavelength dispersion effects cannot be neglected. Further, applications where separate modulations are placed on many, like 10, optical carriers at closely spaced wavelengths that all must propagate down the same fiber in a wavelength multiplexing design, also require sources with greater spectral purity than is possible from a Fabry–Perot laser.

In principle the spectral purity of a Fabry–Perot laser could be improved by passing its output through an external optical filter. But this approach is generally not attractive due to the practical difficulties of maintaining the passband of such a highly selective filter at the same wavelength as the laser. Consequently the method of choice for improved spectral purity is to enhance the wavelength selectivity of the laser cavity by adding an internal filter so that the laser emits on only one longitudinal mode.

An optical grating can provide the desired wavelength selectivity. One way to incorporate a grating into a diode laser cavity is shown in Fig. 2.5(a). As shown by the small arrows in the active layer of this figure, there are small reflections from each of the index changes in the grating. In general each reflection has associated with it a magnitude and phase. Ideally at one wavelength – in practice for a narrow range of wavelengths – all these small reflections add in phase. If the dominant

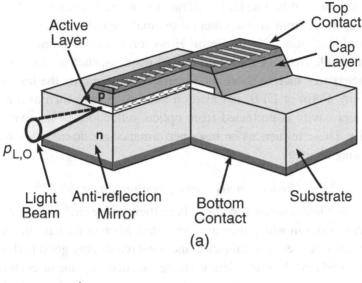

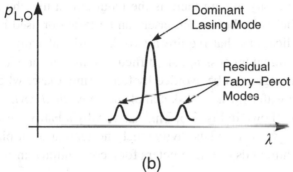

Figure 2.5 (a) Diagram showing the principal components of a distributed feedback (DFB) diode laser with the Bragg grating on top – which is one possible location. (b) Representative spectral output of a DFB diode laser. (Cox, 1996, Fig. 57.2. © 1996 CRC Press, reprinted with permission.)

reflections in the laser cavity come from the wavelength-selective grating, and not the broader band reflections of the cavity end facets, then the laser's spectral character is determined by the grating. The cavity end facet reflections can be suppressed by applying anti-reflection coatings to them. Diode lasers in which the dominant optical feedback necessary for laser operation arises from a grating that is distributed along the cavity are referred to as *distributed feedback*, or DFB, lasers. The typical spectral output of a DFB laser is shown in Fig. 2.5(b); in practice the dominant lasing mode is 20 dB above the residual Fabry–Perot modes.

We now need to derive the incremental modulation efficiency for the DFB laser. Well the DFB laser still has a P vs. I curve like a Fabry–Perot laser, consequently

(2.5) still holds for a DFB. The DFB still has a series resistance, so (2.9) also holds for a DFB. In short, from an incremental or small-signal point of view, the DFB is indistinguishable from a Fabry–Perot! However, this does not imply that a DFB is indistinguishable from a Fabry–Perot in all respects, such as linearity or noise. Laser characteristics such as these are dependent on not only the laser design, in this case Fabry–Perot or DFB, but also on the link design, such as the degree to which the laser cavity is protected from optical reflections of its own light back into its cavity. These influences on laser performance are discussed in more detail in the following chapters.

2.2.1.3 *Vertical cavity surface emitting laser (VCSEL)*

Both of the diode lasers discussed above have their lasing cavity in the plane of the semiconductor wafer in which they are fabricated. Such lasers have the advantages that they are relatively easy to fabricate and have reasonably good performance in terms of threshold and slope efficiency. However, such in-plane lasers have several disadvantages; primary among them is the requirement that the individual lasers be diced from the wafer before the laser can be tested or used in an application. In addition, applications that require a two-dimensional array of lasers, such as interconnects or signal processing, are difficult to implement with in-plane lasers.

These disadvantages can be avoided by fabricating a laser whose cavity is per-pendicular to the wafer surface. A sketch of such a *vertical cavity surface-emitting* (VCSEL) laser is shown in Fig. 2.6. This is basically a Fabry–Perot laser, but with its lasing axis now vertical and its cavity much shorter than an in-plane laser's: a few micrometers vs. hundreds of micrometers for a conventional in-plane Fabry–Perot laser.

Although VCSELs are fabricated in the same semiconductor materials as their in-plane relatives, the design details are quite different. For example, in a VCSEL the photons propagate perpendicular to the active region, rather than in-line with the active region as they do with in-plane lasers. This means that on each pass through the VCSEL cavity there is less chance to generate stimulated photons – i.e. the optical gain is lower. However, the semiconductor materials for VCSELs and in-plane lasers are the same, which means that the losses due to absorption are the same. Thus for the lower gain to still exceed the losses, it is necessary to reduce the emission component of the losses. This can be accomplished by making the mirror reflectivities at either end of the cavity higher, for example $R > 99.8\%$, vs. reflectivities of 65% which are typical for in-plane lasers.

The high reflections at either end of the vertical cavity are provided by multiple, thin layers with alternating dielectric constants. The thicknesses and dielectric con-stants have been chosen to provide high reflection at the bottom, or back, "mirror" and lower reflection at the top, or front, "mirror." The light is confined laterally

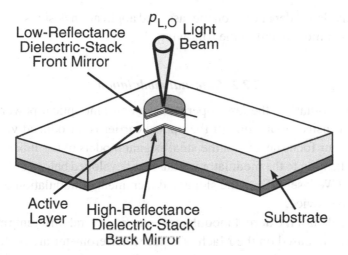

Low-Reflectance
Dielectric-Stack
Front Mirror

$p_{L,O}$ Light
Beam

Active
Layer

High-Reflectance
Dielectric-Stack
Back Mirror

Substrate

Figure 2.6 Diagram showing the principal components of a vertical cavity, surface emitting laser (VCSEL) which is simply a Fabry–Perot diode laser whose laser cavity is perpendicular to the plane of the semiconductor wafer.

within the cavity via either gain or index guiding. One of the ancillary benefits of a vertical cavity laser is that since there is rotational symmetry in the cavity about the vertical axis, the emitted beam is circular. This circular beam – as opposed to the elliptical beam of in-plane lasers – is one of the contributing factors to the increased fiber-coupling efficiency of VCSELs.

The VCSEL cavity is short enough so that only one of its Fabry–Perot modes falls within the semiconductor gain peak. Consequently the VCSEL's emission spectrum has only one line, like a DFB, but without the need for the frequency selective DFB grating. As a benchmark, a "high" optical power from a VCSEL operating in a single mode is 4.8 mW (Choquette and Hou, 1997). However, there are typically several transverse cavity modes and when the VCSEL is operated with multiple transverse modes the total optical output power can be greater than 100 mW. In addition to their effect on total output power, the cavity modes also have an impact on the VCSEL-to-fiber coupling efficiency. Higher efficiency is possible under the multi-mode condition, but the fiber coupling efficiency is much more sensitive to mis-alignment and the optical spectrum no longer consists of a single frequency.

Once again we need to derive the incremental modulation efficiency for a VCSEL. Well the VCSEL still has a P vs. I curve like a Fabry–Perot laser, consequently (2.5) still holds for a VCSEL. The VCSEL still has a series resistance, so (2.9) also holds for a VCSEL. In short, from an incremental or small-signal point of view, the VCSEL is also indistinguishable from a Fabry–Perot! However, as was the case with the DFB laser, the noise and linearity properties of the VCSEL may be different from the Fabry–Perot depending on the laser and application designs. These aspects are discussed in more detail in the following chapters.

2.2.2 External modulation

For external modulation, the laser operates at a constant optical power (CW) and the desired intensity modulation of the optical carrier is introduced via a separate device. Since our focus here is on the small-signal models of the link components, we turn our attention to the modulator because, as developed below, only the optical power of the CW laser affects the slope and incremental modulation efficiency of the modulation device.

Like the diode laser, external modulators come in several different implementations. Modulators based on the Mach–Zehnder interferometer are by far the most common; however, modulators based on the directional coupler or the electro-absorption effect in semiconductors have been improving in performance and hence are increasing in popularity. Again since our goal in this chapter is to develop the basic device models upon which we can build the link design parameters in future chapters, we focus here on modeling the Mach–Zehnder modulator, both because of its popularity and because of its analytic tractability. Models for any other type of modulator can be developed using this same methodology; however, we only do so here for the directional coupler and the electro-absorption modulators.

The key material property for making a modulator is that a change in an electrical parameter must alter an optical property of the material. In lithium niobate, for example, an applied electric field alters the optical index of refraction. Other material properties that are also important are optical loss, maximum optical power and stability – both thermal and optical. Of the three classes of materials from which a modulator can be made: inorganics, semiconductors and electro-optic polymers, by far the most common modulator material at present is the inorganic lithium niobate. This material has probably succeeded not because it is the best with respect to loss, stability, maximum optical power or electro-optic sensitivity, but because it has a reasonable combination of all four. It is also relatively inexpensive, since this material was developed for and is widely used in surface acoustic wave (SAW) filters.

2.2.2.1 Mach–Zehnder modulator

Early "bulk" modulators simply placed electrodes on a block of lithium niobate. The resulting electric field alters the index of refraction, which in turn alters the phase of an optical wave passing through the lithium niobate. Orders of magnitude greater sensitivity, as expressed as optical phase shift per volt of applied field, are possible by confining the light to an optical waveguide and placing electrodes alongside

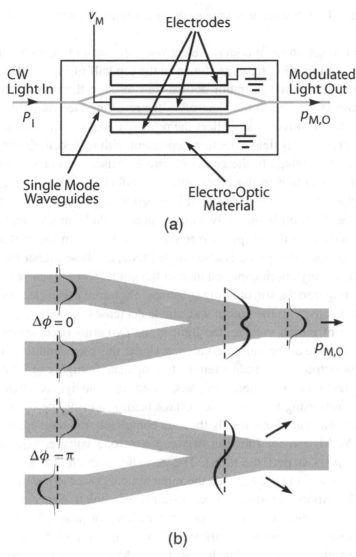

Figure 2.7 (a) Diagram showing the layout of an optical intensity modulator based on a Mach–Zehnder interferometer. (b) Detail showing wavefronts at the output combiner for 0 and 180 degree phase shifts between the waves. (Cox, 1996, Fig. 57.14. © 1996 CRC Press, reprinted with permission.)

(Alferness, 1982). By imbedding such an optical phase modulator in an interferometer, it is possible to convert the phase modulation to intensity modulation. And the interferometer of almost universal choice is the Mach–Zehnder.

The layout of a typical Mach–Zehnder modulator, MZM, (Martin, 1975) is shown in Fig. 2.7(a). The waveguides all propagate a single spatial optical mode. The incoming CW light from the laser is typically split equally and fed into two nominally equal length arms, whose outputs recombine to feed the modulator output waveguide.

With zero voltage on the modulator electrodes, the light in the two arms arrives at the output combiner in phase, as shown in the top half of Fig. 2.7(b). Thus this bias condition results in maximum optical transmission through the modulator. When a voltage is applied to the electrodes, the resulting electric field – which is perpendicular to the waveguide – alters the refractive index in the two arms. Since the electro-optic effect is field-direction dependent with respect to the propagation axes, applying the voltage to the center electrode causes the index to increase in one arm and to decrease in the other arm. The effect of these index changes is the introduction of a relative phase shift between the light in the two arms when they recombine. The combining of two out-of-phase light beams attempts to excite higher order modes in the output waveguide, as indicated in the bottom half of Fig. 2.7(b). But since this guide is also single mode, all these higher order modes cannot propagate any significant distance in the guide. Consequently this light is lost via scattering into the surrounding substrate. Therefore as the electrode voltage increases, the intensity in the output waveguide decreases.

For a sufficiently high voltage, the light in the two arms becomes exactly 180 degrees out of phase at the output combiner. Under this bias condition, ideally all the light in the output waveguide attempts to propagate in higher order modes and none propagates in the fundamental mode. Consequently no light reaches the output fiber. Further increasing the electrode voltage begins to bring the light in the two arms closer in phase until eventually they are completely in phase again.

Thus the MZM transfer function between electrode voltage, v_M, and optical output power, $p_{M,O}$, is periodic in v_M. The periodic function is a raised cosine as shown by the curve in Fig. 2.8, when the input optical power to the modulator, P_I, is constant. An important modulator parameter from a link design standpoint is the voltage required to induce 180 degrees, or π radians, of phase shift between the two optical waves at the output combiner; this voltage is referred to as V_π.

Even at the peaks of transmission through the MZM there will be some loss, which in lithium niobate modulators is typically 3 to 5 dB and 10 to 12 dB in polymer and semiconductor modulators. In lithium niobate modulators the dominant loss component – typically 2 to 4 dB – is the fiber–waveguide coupling, whereas in the other material systems the waveguide loss usually dominates. Regardless of the source of the loss, it has the same effect on the link design. Consequently for our purposes we can lump all the losses into a single term, which we denote T_{FF} for *transmission fiber to fiber*; a representative value of 0.5, or 3 dB, is shown in Fig. 2.8.

As we did for the diode laser, we want to obtain the slope efficiency and incremental modulation efficiency for the MZM. However, in contrast to the previous

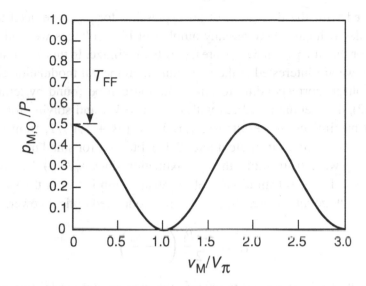

Figure 2.8 Mach–Zehnder modulator optical transfer function $p_{M,O}/P_I$ versus v_M/V_π, for the case where $T_{FF} = 0.5$.

development, we formulate these parameters in the reverse order from the diode laser case, primarily because the slope efficiency for an external modulator is a derived parameter, not an intrinsic parameter of the MZM, as it was with the diode laser.

To formulate the incremental modulation efficiency, we begin with the functional form of the MZM transfer function plotted in Fig. 2.8:

$$p_{M,O} = \frac{T_{FF}P_I}{2}\left[1 + \cos\left(\frac{\pi v_M}{V_\pi}\right)\right]. \tag{2.12}$$

Here, T_{FF} is the ratio $p_{M,O}/P_I$ when the interferometer is biased for maximum optical transmission. For the modulator transfer function shown in Fig. 2.8, T_{FF} is measured at $v_M = 2kV_\pi$, where k is an integer. When an electrical modulation voltage, v_m, is applied to the modulator in addition to a dc bias voltage, V_M, the total instantaneous modulator voltage can be written as $v_M = V_M + v_m$. Substituting this expression for v_M into (2.12) and expanding the cosine sum yields

$$p_{M,O} = \frac{T_{FF}P_I}{2}\left[1 + \cos\left(\frac{\pi V_M}{V_\pi}\right)\cos\left(\frac{\pi v_m}{V_\pi}\right) - \sin\left(\frac{\pi V_M}{V_\pi}\right)\sin\left(\frac{\pi v_m}{V_\pi}\right)\right]. \tag{2.13}$$

From (2.13) we can identify the dc bias term, $P_{M,O} = T_{FF}P_I/2$, and the modulation terms, $p_{m,o}$, which are the second and third terms in (2.13).

To proceed with the derivation of an expression for $p_{m,o}$, we need to choose a specific bias voltage. There are any number of bias voltages we could choose, depending on the link parameter we are trying to maximize. In this discussion, let us assume that we are interested in the maximum incremental modulation efficiency. The bias voltage corresponding to this condition can be found by letting $v_M = V_M$ in (2.12), differentiating (2.12) with respect to V_M and solving for the value of V_M that maximizes the result. Answer: $V_M = (2k + 1)V_\pi/2$, from which we choose $V_M = V_\pi/2$ and substitute it into (2.13). Further, for small modulation, i.e. for $v_m \ll V_M$, we can substitute the approximation $\sin(\pi v_m/V_\pi) \cong \pi v_m/V_\pi$ into (2.13) as well. These substitutions yield a relationship between the small-signal modulation voltage and the resulting intensity modulated optical power:

$$p_{m,o} \cong \frac{T_{FF} P_I}{2} \left(-\frac{\pi v_m}{V_\pi} \right). \qquad (2.14)$$

This is half of what we need to express the incremental modulation efficiency. The other half requires us to have the low frequency, lumped-element electrical circuit model for a MZM. At low frequencies – typically less than 1 GHz – the modulator electrodes form a capacitor, C_M. However, as we did with the analogous development for the diode laser, we neglect the frequency dependent components in this section. With this assumption, what remains of the modulator impedance is the resistance, R_M, which represents the RF losses of the modulator electrodes and dielectric. We can model these RF losses either as a series or a shunt resistor; we choose the latter. Since these losses are small, the value of R_M is greater than the source resistance. Consequently to establish a match with the source resistance simply requires us to place a matching resistor, R_{MATCH} in parallel with R_M where the value of R_{MATCH} is chosen such that the parallel combination equals the source resistance, R_S.

Combining this low frequency approximation of the modulator impedance with the modulation source model we used with the diode laser, we obtain the circuit shown in Fig. 2.9. The relation between v_s and v_m can be obtained by inspection of the resistive divider: $v_m = v_s/2$. Substituting this value for v_m into the expression for available power from the source, (2.7), yields the available power in terms of v_m:

$$p_{s,a} = \frac{v_s^2}{4R_S} = \frac{v_m^2}{R_S}. \qquad (2.15)$$

Squaring (2.14) and substituting (2.15) into the result enables us to eliminate the intermediate variable v_m. The result of these steps produces an expression for the low frequency, incremental modulation efficiency of a MZM with a parallel matching

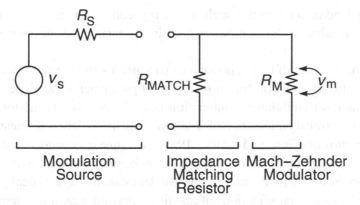

Figure 2.9 Low-frequency approximation of a Mach-Zehnder modulator and impedance matching resistor fed by a modulation source.

resistor:

$$\frac{p_{m,o}^2}{p_{s,a}} = \left(\frac{T_{FF}\,P_I\pi}{2V_\pi}\right)^2 R_S. \tag{2.16}$$

Now that we have the incremental modulation efficiency for the MZM, getting its slope efficiency is easy. Basically we want to group the terms in (2.16) so that they have the same format as (2.11). This can be done by simply moving R_S into the squared term and then dividing the result by R_S:

$$\frac{p_{m,o}^2}{p_{s,a}} = \left(\frac{T_{FF}\,P_I\pi R_S}{2V_\pi}\right)^2 \frac{1}{R_S}. \tag{2.17}$$

The squared term with R_S included now has the units of $(W/A)^2$ which are the units of slope efficiency squared. Thus (2.17) has the same form as (2.11) if we denote MZM fiber-coupled slope efficiency, s_{mz}, as

$$s_{mz}\left(V_M = \frac{V_\pi}{2}\right) = \frac{T_{FF}\,P_I\pi R_S}{2V_\pi}. \tag{2.18}$$

We have shown the explicit dependence of slope efficiency on bias point as a reminder that the expression shown in (2.18) is valid only at a particular bias point. An analogous derivation would be needed to calculate the slope efficiency at another bias point.

As we did with the diode laser, it is instructive to ask: what is the maximum slope efficiency for the MZM? To answer this question note that all the parameters in (2.18) can be selected independently and that each parameter is only limited by practical considerations such as how much optical power the waveguides can handle, not fundamental limits such as energy conservation. Consequently the answer to the maximum slope efficiency question is that in principle there is no upper bound!

An important aspect of (2.18) is that a *dc* parameter – the average optical power, P_I – enters into the equation for an *incremental* parameter – the slope efficiency, s_{mz}. Thus the laser/modulator combination is analogous to a transistor in which the *incremental* transconductance of a bipolar transistor depends linearly on the *dc* collector current (Gray and Searle, 1969). The linear dependence of external modulation slope efficiency on optical power is in sharp contrast with the virtual independence of direct modulation slope efficiency on average optical power. We will see the important ramifications of this linear dependence in Chapters 3 and 4.

The wavelength dependence of s_{mz} takes two different forms depending on precisely what we are comparing (Ackerman, 1997). One form of the wavelength dependence arises if we compare V_π, and hence s_{mz}, for a fixed modulator design at two different wavelengths. In this case the phase shift in the optical wave for a given modulator electrode length is inversely proportional to the wavelength. Consequently for a given modulator operating at two different wavelengths, s_{mz} scales as λ^{-1}.

The other form of the wavelength dependence for a Mach–Zehnder arises when comparing the s_{mz} of two modulators, each of which has been optimized for operation at a different wavelength. In this case a second wavelength dependence comes into play, in addition to the one just discussed. The shorter wavelength permits a smaller single mode optical waveguide, thereby permitting a narrower gap between the modulation electrodes and consequently increasing the electrical field across the optical waveguide. This effect scales approximately inversely as wavelength. Consequently for two modulators, each of which was optimized for a different wavelength, s_{mz} scales as approximately λ^{-2}.

To get a feel for what some typical slope efficiencies for a MZM are, we plot s_{mz} versus wavelength for a range of CW optical powers and modulator V_πs that are typical of those in use. Commercially available MZMs fabricated in lithium niobate generally have V_πs in the range of 15 V down to 2 V, with 0.35 V being the lowest reported (Betts *et al.*, 1988). Early modulators, especially those operating at 0.85 μm, used low optical powers, often less than 1 mW. More recently modulators have been designed to operate at 1.3 μm with CW optical powers around 100 mW, with 400 mW being the highest reported (Cox *et al.*, 1996). From (2.18) it is clear that it is not the individual values of P_I and V_π that are important but rather the ratio of P_I/V_π. Consequently combining the extremes for P_I and V_π given here and rounding to the nearest order of magnitude produces ratios of P_I/V_π in the range of 0.0001 to 1 W/V. The values of s_{mz} corresponding to these ratios of P_I/V_π are plotted in Fig. 2.10, along with some experimentally reported values.

The corresponding bandwidths for each of the devices in Fig. 2.10 are presented in Chapter 4. In addition to data on Mach–Zehnder modulators, Fig. 2.10 also includes data on directional coupler and electro-absorption modulators, which are discussed in Sections 2.2.2.2 and 2.2.2.3, respectively.

In comparing Figs. 2.3 and 2.10, we see that both in theory and in practice it is presently possible to achieve higher slope efficiencies with a MZM fed by a high power laser than it is with direct modulation of a diode laser.

2.2.2.2 Directional coupler modulator

A modulator similar to the MZM is based on a *directional coupler* (Papuchon *et al.*, 1977). In a standard passive coupler, two waveguides are run parallel and close enough to each other that evanescent coupling can occur between them. The length of the coupling region is selected so that the desired degree of power is coupled from one guide to the other.

To make a modulator from this passive coupler simply requires placing electrodes parallel to the waveguides, as shown in Fig. 2.11(a) and fabricating this whole structure in an electro-optic material, such as lithium niobate. With no voltage applied to the electrodes, the waveguide spacing and separation are chosen such that, ideally, there is complete transfer between the waveguides. Consequently all the optical power in one of the incident guides transfers to the other output guide. As the voltage on the electrodes is increased, the resulting electric field alters the refractive index, which in turn changes the effective coupling length, thereby permitting some of the optical power in the input guide to stay in that guide to the output. For a sufficiently high electrode voltage, all the optical power in the incident waveguide will, ideally, remain in that guide to the output. The voltage needed to go from full coupling to zero coupling is called the switching voltage, V_S.

It can be shown (Halemane and Korotky, 1990) that the transfer function between the electrode voltage, v_M and the optical output, $p_{D,O}$ for a directional coupler modulator, DCM, is given by

$$p_{D,O} = T_{FF} P_I \frac{\sin^2\left[\dfrac{\pi}{2}\sqrt{1 + 3\left(\dfrac{v_M}{V_S}\right)^2}\right]}{1 + 3\left(\dfrac{v_M}{V_S}\right)^2}, \qquad (2.19)$$

where P_I is the optical power in the input waveguide. Although V_S for the DCM plays an analogous role to V_π for the MZM, the DCM transfer function – unlike that for the MZM – is not periodic in V_S. This is readily apparent from (2.19) or from the plot of this equation shown in Fig. 2.11(b).

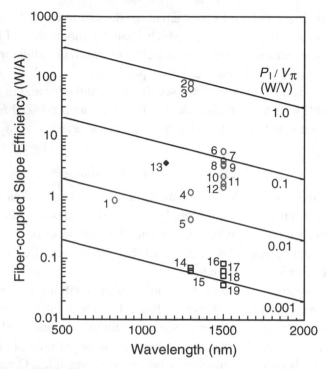

Figure 2.10 Plot of representative external modulator fiber-coupled slope effi-ciency vs. wavelength for the ratio P_I/V_π between 0.001 and 1 W/V, along with data points for recently reported results for three types of modulators. ○ MZ LiNbO$_3$, □ electro-absorption, ◆ directional coupler.

Point	Authors and date	Frequency (GHz) dBe	Point	Authors and date	Frequency (GHz) dBe
1	C. Gee *et al.*, 1993	17	11	K. Noguchi *et al.*,	20
2	C. Cox *et al.*, 1996	0.15		1991	
3	G. Betts *et al.*, 1989	0.15	12	K. Noguchi *et al.*,	50
4	R. Jungerman and D.	32		1994	
	Dolfi, 1992		13	O. Mikami *et al.*, 1978	1
5	A. Wey *et al.*, 1987	18	14	C. Rolland *et al.*, 1991	11
6	K. Kawano *et al.*,	5	15	Y. Liu *et al.*, 1994	20
	1991		16	T. Ido *et al.*, 1994	2
7	R. Madabhushi, 1996	18	17	I. Kotaka *et al.*, 1991	16.2
8	K. Noguchi *et al.*,	12	18	T. Ido *et al.*, 1995	not
	1993				reported
9	K. Kawano *et al.*,	14	19	F. Devaux *et al.*, 1994	36
	1989				
10	M. Rangaraj	5			
	1992				

Assumptions:

The optical power into Mach–Zehnder and directional coupler modulators was 400 mW, which at present is about the maximum for lithium niobate-based devices. The optical power into EA modulator was assumed to be 10 mW, which is representative of the state of the art maximum for this type of modulator. Finally the RF impedance given in the papers was used.

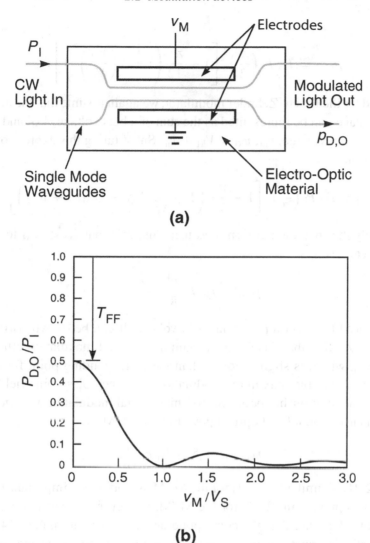

(a)

(b)

Figure 2.11 (a) Diagram showing the layout of an optical intensity modulator based on a directional coupler. (b) Directional coupler modulator optical transfer function $P_{D,O}/P_I$ vs. v_M/V_S for the case where $T_{FF} = 0.5$. (Cox, 1996, Fig. 57.16. © 1996 CRC Press, reprinted with permission.)

To formulate the incremental slope efficiency for the DCM, we begin by representing the $\sin^2$ term in a power series (Gradshteyn and Ryzhik, 1965):

$$\sin^2 x = \sum_{1}^{\infty}(-1)^{k+1}\frac{2^{2k-1}x^{2k}}{(2k)!}. \tag{2.20}$$

For the slope efficiency, we only need the first two terms from (2.20), i.e. $k = 2$, to represent the $\sin^2$ term of (2.19). After some algebraic simplification we get

$$p_{D,O} \cong T_{FF} P_I \left(\frac{\pi}{2}\right)^2 \left[1 - \frac{\pi^2}{12}\left(1 + 3\left(\frac{v_M}{V_S}\right)^2\right)\right]. \qquad (2.21)$$

As we did for the Mach–Zehnder modulator, we again assume that the modulator electrode voltage can be represented as the sum of a bias voltage, V_M, and a small-signal modulation voltage, v_m; $v_M = V_M + v_m$. Substituting this expression for v_M into (2.21) yields

$$p_{D,O} = T_{FF} P_I \left(\frac{\pi}{2}\right)^2 \left[1 - \frac{\pi^2}{12}\left(1 + \frac{3}{V_S^2}\left(V_M^2 + 2V_M v_m + v_m^2\right)\right)\right]. \qquad (2.22)$$

Keeping only the incremental term and using the fact that $v_m \ll V_M$ to simplify (2.22) we get

$$p_{d,o} = T_{FF} P_I \frac{\pi^4}{8} \frac{V_M v_m}{V_S^2}. \qquad (2.23)$$

We now need to select a particular bias voltage. It can be shown (Bridges and Schaffner, 1995) that the bias for maximum incremental modulation efficiency is $V_M \cong 0.43 V_S$, which is slightly lower than the corresponding point for a MZM. Evaluating (2.23) at the maximum modulation efficiency bias point yields an expression for the relationship between the small-signal modulation voltage and the resulting intensity modulated optical power of the DCM:

$$p_{d,o,\,max} = 5.24 T_{FF} P_I \frac{v_m}{V_S}. \qquad (2.24)$$

Equation (2.24) is similar to (2.14) for the MZM, including the important P_I dependence that was present in (2.14) for the MZM. However, the constant term, 5.24, is a factor of 3.3 greater than the corresponding $\pi/2 = 1.57$ term in (2.14). We do not realize all this improvement because the common electrode arrangement with a MZM – where the center electrode is driven – results in a push-pull configuration whereas with the DCM only single ended drive is possible because there is no center electrode. Thus for electrodes of *equal* length, the MZM V_π is a factor of 2 less than the DCM V_S. Consequently the net increase of the DCM incremental modulation efficiency over the MZM is 3.3/2 or about a factor of 1.6.

To obtain the DCM incremental modulation efficiency, we need the low frequency, lumped-element electrical circuit model for the DCM electrodes. The DCM modulator electrodes form a capacitor, just like the MZM electrodes did. Therefore if we terminate the DCM electrode capacitance with a resistor, as we did for the MZM, we can use the same derivation in getting to (2.15) for the DCM as we did for the MZM. Thus we also have the same circuit for the DCM as we did for the MZM as shown in Fig. 2.15. Consequently squaring (2.24) and combining the result with (2.15) to eliminate the intermediate variable v_m results in the expression for the low frequency, incremental modulation efficiency of a DCM with a parallel matching resistor:

$$\frac{p_{d,o}^2}{p_{s,a}} = \left(\frac{5.24 T_{FF} P_I}{V_S}\right)^2 R_S. \tag{2.25}$$

The DCM slope efficiency, s_{dc}, follows from (2.25) using the same procedure we used in transforming (2.16) into (2.17):

$$s_{dc}\left(V_M \cong 0.43 V_S\right) = \frac{5.24 T_{FF} P_I R_S}{V_S}, \tag{2.26}$$

where again we have included explicit reference to the bias point where this slope efficiency was calculated.

The functional form of (2.26) is the same as the corresponding MZM expression, (2.18). Thus we do not repeat the MZM plot of typical slope efficiencies for the DCM.

The slightly higher slope efficiency for the DCM is often offset in practice by its more demanding fabrication, especially if long electrode lengths are required. This is because increasing the length of a MZM is simply a matter of making the waveguides in the two arms longer but does not affect the spacing *between* the waveguides in the two arms. Increasing the length of a DCM involves recalculating the waveguide coupling spacing as well as increasing the electrode length. Consequently the directional coupler is favored over the MZM when a compact 2×2 switch is desired, but the MZM is favored when attempting to maximize slope efficiency.

2.2.2.3 Electro-absorption modulator

Although some semiconductors do have a linear electro-optic coefficient, by far the more common semiconductor-based optical modulators use the electro-absorption, EA, effect. To appreciate the basis for this modulation approach, we begin with a three layer semiconductor structure consisting of two oppositely doped layers – p and n – separated by an undoped or intrinsic, I, layer. Such a PIN structure exhibits an abrupt change in its absorption as a function of wavelength; the wavelength of this abrupt change is referred to as the bandgap.

The change in absorption around the bandgap is often portrayed as a step function; i.e. the absorption is zero for wavelengths longer than the bandgap and infinite for wavelengths shorter than the bandgap. Actually the change has a finite slope as shown in Fig. 2.12 (Knupfer *et al.*, 1993) that contains the calculated absorption vs. wavelength for a PIN structure, fabricated in gallium arsenide. With no voltage

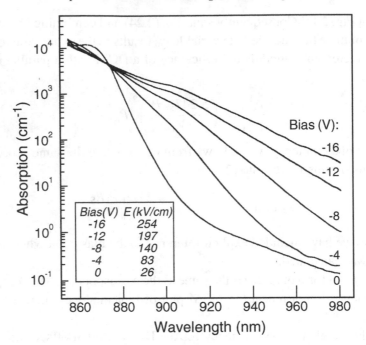

Figure 2.12 Calculated absorption coefficient vs. wavelength with electric field as the parameter for a PIN junction fabricated in GaAs (Knupfer *et al.*, © 1993 IEEE, reprinted with permission.)

applied across the junction, we see the "bandgap" where the absorption changes by over four orders of magnitude with just a 40 nm change in wavelength.[3]

As the bias across the semiconductor junction is increased from zero, the shape of the absorption vs. wavelength function changes; see again Fig. 2.12. It is common to refer to this change as a "shift" in the bandgap energy with bias. However, in fact it is more accurate to think of the bandgap as remaining at a fixed wavelength and the change arising from the addition of an exponential "tail" to the absorption function that increases with increases in the bias applied across the junction. When such a bias dependent change in the absorption function occurs in a bulk semiconductor, it is referred to as the Franz–Keldysh effect. When a similar effect occurs in quantum well structures, it is referred to as the quantum confined Stark effect, QCSE.

To use the Franz–Keldysh effect or QCSE for modulation, the wavelength of the optical carrier to be modulated is chosen such that the semiconductor is almost transparent at zero applied electric field. Consequently this wavelength is increasingly attenuated as the electric field is increased. Thus in contrast to the

---

[3] As a point of reference for how abrupt a change this is, over the approximately 250 nm wavelength span of the visual spectrum, the human eye/brain can discriminate wavelength differences of 1–2 nm (see for example Graham *et al.*, 1965).

modulators presented above, an EA modulator requires some degree of wavelength stabilization, or at least wavelength coordination, between it and the CW source.

An EA modulator can be fabricated in a number of configurations. One of the common ones is virtually identical to the Fabry–Perot semiconductor laser shown in Fig. 2.1(a). The principal difference is that unlike a laser, the ends of an EA modulator are anti-reflection coated to reduce the fiber-coupling loss. Also in an EA modulator, the p-n-junction is reverse biased – vs. the laser's forward bias – and the doping in an EA modulator has been designed to increase the electrical field dependent, and hence voltage dependent, absorption. Unlike modulators fabricated in lithium niobate, semiconductor-based EA modulators can be monolithically integrated with the CW laser. This integration eliminates two of the three expensive and optically lossy fiber couplings required without integration.

For the EA modulator, we derive a general expression for the slope and incremental modulation efficiencies as a function of bias voltage, rather than deriving them for a specific bias voltage – as we did for the other types of modulation devices. We begin with the expression that relates the voltage dependent absorption, $\Delta\alpha(v_M)$, to the optical power passing through the modulator, $p_{A,O}$ (Welstand, 1997):

$$p_{A,O} = T_{FF} P_I \, e^{-\Gamma L \Delta\alpha(v_M)}, \qquad (2.27)$$

where Γ is the optical confinement factor, or percentage of optical power in the waveguide that is confined to the absorbing layer, L is the waveguide length and the other symbols are as previously defined. Note that we used $\Delta\alpha(v_M)$ in (2.27), which is the normalized absorption coefficient in the sense that $\Delta\alpha(0) = 0|_{v_M - 0}$. The expression for the absorption as a function of voltage is complex and adds little insight to the link design process. Thus we do not present it here, but for completeness include it in Appendix 2.2.

Next we express the modulator voltage as the sum of dc and incremental terms, viz. $v_M = V_M + v_m$, and substitute this into (2.27):

$$p_{A,O} = T_{FF} P_I \, e^{-\Gamma L \Delta\alpha(V_M) - \Gamma L \Delta\alpha(v_m)}. \qquad (2.28)$$

The first exponential represents the absorption due to the dc bias voltage; we denote this term by T_N:

$$T_N = e^{-\Gamma L \Delta\alpha(V_M)}. \qquad (2.29)$$

For small modulation, we can use the approximation $e^{-\Gamma L \Delta\alpha(v_m)} \cong 1 - \Gamma L \Delta\alpha(v_m)$. Making these two substitutions in (2.28) and differentiating the result with respect to v_m yields an expression for the small-signal modulated optical power:

$$p_{a,o} = \left.\frac{dp_{A,O}}{dv_m}\right|_{V_M} = -T_{FF}P_IT_N\Gamma L\left(\left.\frac{d\Delta\alpha(v_m)}{dv_m}\right|_{V_M}\right)v_m. \tag{2.30}$$

To put (2.30) into a form that is analogous to the expressions for the MZM and DCM, we introduce a voltage V_α, which is analogous to V_π:

$$V_\alpha = \frac{-1}{\Gamma L\left(\left.\dfrac{d\Delta\alpha(v_m)}{dv_m}\right|_{V_M}\right)}. \tag{2.31}$$

Substituting (2.31) into (2.30) we obtain the desired relationship between the modulation voltage and the modulated optical power:

$$p_{a,o} \cong T_{FF}P_IT_N\frac{v_m}{V_\alpha}. \tag{2.32}$$

We can finish developing the EA incremental modulation efficiency by specifying it in a particular circuit. A simple one to use is a resistive termination, $R_{MATCH} = R_S$, in parallel with the EA modulator. In doing so we arrive at a circuit identical to the one used for the previous two modulators as shown in Fig. 2.15. Squaring (2.32) and substituting (2.15) into the result yields

$$\frac{p_{a,o}^2}{p_{s,a}} = \left(\frac{T_{FF}P_IT_N}{V_\alpha}\right)^2 R_S = \frac{s_{ea}^2}{R_S}, \tag{2.33}$$

where the EA slope efficiency, s_{ea} is defined as

$$s_{ea}(V_M) = \frac{T_{FF}P_IT_N(V_M)R_S}{V_\alpha(V_M)}. \tag{2.34}$$

To calculate the slope efficiency for an EA modulator, we would need to select a specific bias point, V_M, and then obtain specific values for T_N by evaluating (2.29), and for V_α by evaluating (2.31). Thus unlike the previous slope efficiency expressions, which were only valid at one bias point, (2.34) is a general expression for EA modulator slope efficiency.

The functional form of (2.34) is the same as (2.18) for the MZM and (2.26) for the DCM. Thus we once again have a modulator where the slope efficiency is, at least in principle, unbounded. However, the practical difficulties of using an EA modulator at high optical powers center on the fact that the unmodulated optical power is *absorbed* in these modulators, not *scattered* into the substrate as it was in the MZM or *coupled* into another waveguide as it was in the DCM. As a result of this fact, the maximum optical power for EA modulators is presently about 40 mW, or about an order of magnitude lower than the current maximum optical power for MZMs in lithium niobate.

EA modulators, unlike either of the other modulators we have discussed, also require that the wavelength of the CW optical source be within about 40 nm of its absorption edge. If a diode laser is used as the CW source, this can readily be accomplished via temperature control of the laser. Alternatively since EA modulators are fabricated in semiconductors, the diode laser can be fabricated on the same substrate much more readily than with lithium niobate modulators. Such integration of laser and modulator also eases the wavelength tolerance, since both track more closely with temperature.

2.3 Photodetectors

Whether we use direct or external modulation, the output of all the above modulation devices is an intensity modulated optical wave. To recover the modulation from this optical carrier requires: (1) a material that can absorb the optical wave; (2) a device structure that can be fabricated in this material; and (3) the ability to collect the carriers generated by the absorption and to convey them to an external circuit. Lots of materials absorb light in the wavelength range of interest here, so the fact that semiconductors are the universal material of choice is based on their ability to perform the latter two functions. As with the diode lasers and external modulators, there are numerous device designs that are capable of performing these functions (see for example Yu, 1997). For the purposes of this chapter, we choose to focus on the p-intrinsic-n (PIN) photodiode, which is by far the most common photodetector in use for communication applications at the present time.

Figure 2.13 shows the basic layout of a PIN photodiode. The intrinsic or "I" layer is an undoped layer within the semiconductor where the optical absorption occurs. The regularity of the semiconductor crystal structure imposes discrete or quantized energy levels within the crystal. Consequently no absorption occurs for photons with energy less than the minimum energy between quantized levels. Thus for photons whose energy is less than the bandgap energy, the semiconductor is transparent. In principle, all the photons whose energy is greater than the bandgap energy are absorbed. However, other design constraints, such as the thickness of the I-layer, usually limit the maximum absorption that is achieved in practice to less than 100%.

The bandgap energy for silicon corresponds to a wavelength near 1.1 μm, so this material is suitable for the fabrication of photodetectors for links operating around the 0.85 μm wavelength band. For the 1.3 and 1.55 μm wavelength bands, germanium and the compound semiconductor indium-gallium-arsenide have bandgap energies that correspond to wavelengths around 1.6 μm. Consequently both of these

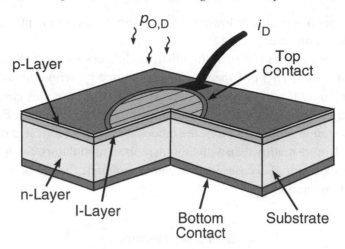

Figure 2.13 Sketch showing the principal components of a low frequency, positive-intrinsic-negative (PIN) photodetector with front surface illumination.

materials are used to make photodetectors for the long wavelength bands, although germanium is used less at present because of its typically three orders of magnitude higher dark current, i.e. the current that leaks through an unilluminated photodiode.

Absorption of the electrically neutral photon results in the creation of a pair of oppositely charged electrical carriers, so charge neutrality is preserved. In the absence of any electric field, these photo-generated carriers would drift around within the I-layer and eventually non-radiatively recombine, i.e. recombine without emitting a photon. Therefore, to collect these photo-generated carriers, an electric field is set up across the I-layer. A convenient way to establish such a field is to place oppositely doped layers on either side of the I-layer. The n-layer is doped so that an excess of negative electron carriers is available, while the p-layer is doped with an excess of positive hole carriers. The electric field that results from these excess carrier densities sweeps the photo-generated carriers out of the I-layer. By placing contacts on the p- and n-layers, the collected photo-generated carriers can be conveyed to an external circuit.

The alert reader will note that since doping has only a second-order effect on the bandgap energy, photons are absorbed in the p- and n-layers as well. But since there is no field in these layers, only those photo-generated carriers that drift or diffuse into the I-layer contribute to the photocurrent; the remainder will non-radiatively recombine. In a common PIN structure, the incident optical wave has to pass through either the p- or n-layer to reach the I-layer. Therefore one of the tradeoffs in photodetector design is to minimize the thickness of the top doped layer, without incurring the deleterious effects that a thin doped layer has on the series resistance of that layer.

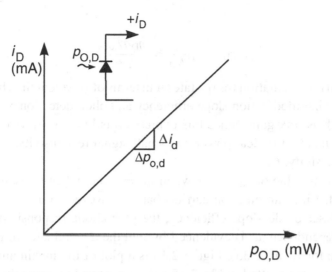

Figure 2.14 Photo-generated current vs. incident optical power response curve for a reverse biased PIN photodiode.

The photodetector structure that we evolved above is, from an electrical point of view, a diode. Consequently forward biasing such a structure would cause a large electrical current to flow that would, in general, be much larger than the photo-generated current. At zero bias, no electrical bias current would flow. However, zero bias operation is typically not used because the I-layer is generally not fully depleted of carriers, which, as we will see below, leads to increased distortion and lower frequency response. Thus PIN photodiodes are usually operated with at least a few volts of reverse bias applied to them. Under reverse bias, the current vs. optical power "curve" for a PIN photodiode is the straight line shown in Fig. 2.14. The slope of this line is the *responsivity* of the photodiode; its units are A/W.

The photodiode responsivity, r_D, can be expressed in quite straightforward terms; it is simply the product of the external quantum efficiency, η_D, and the ratio of the electron charge to the photon energy. This relationship is shown after the first equals sign in (2.35) below.

In turn η_D is defined as the ratio of the number of photo-generated electrons to the number of incident photons. By designating the external quantum efficiency, this includes all the losses associated with the photon-to-electron conversion process. Thus η_D includes optical losses, such as reflection at the photodiode surface; electro-optical losses, such as absorption outside of the I-layer; and electrical losses, such as carrier recombination before collection by the doped regions and internal efficiency, i.e. factors such as whether the semiconductor has a direct or indirect bandgap. Collecting the above terms yields an expression for the photodiode responsivity:

$$r_D = \eta_D \frac{q}{h\nu} = \frac{\eta_D q \lambda_0}{hc}. \tag{2.35}$$

The right hand representation for r_D states it in terms of wavelength, which is useful when we combine modulation slope efficiency and photodetection responsivity in equations such as link gain. Since this expression is linear over several orders of magnitude in incident optical power, the small-signal responsivity, r_d, is equal to the total responsivity, r_D.

The maximum value of r_d occurs when $\eta_D = 1$, i.e. when every incident photon is converted into an electron carrier that is conveyed to the external circuit. As with the laser diode slope efficiency, the photodiode responsivity also has a wavelength dependence, as is evidenced by ν in the denominator of the left hand representation for r_D in (2.35). Figure 2.15 is a plot of the maximum value of r_d vs. wavelength. Also plotted in Fig. 2.15 are the reported responsivities of several research and commercially available, fiber-coupled photodiodes. The corresponding bandwidths for each of the devices in Fig. 2.15 are presented in Chapter 4. As we can see from the data in Fig. 2.15, the actual photodiodes come much closer to the theoretical maximum responsivity than do diode lasers to the corresponding theoretical maximum slope efficiency. The primary reason for the higher r_d in practical devices, at least for low frequency photodiodes, is that the fiber coupling is not into a waveguide, as it was with the laser, but rather into the photodiode I-layer, which has a diameter that is typically as large as – and often larger than – the fiber core diameter. At higher frequencies a waveguide structure is used for the photodiode absorbing region which introduces the same coupling inefficiency as the laser.

For link calculations, we need the incremental detection efficiency for the photodiode. We begin this derivation with the small-signal, equivalent circuit model for the PIN photodiode. Since the optical-to-electrical conversion process depends only weakly on the photodiode bias voltage, we can represent this conversion by an ideal current source. The conversion constant between incident optical power, p_{od}, and the detected electrical current, i_d, is the responsivity; thus

$$i_d = r_d p_{od}. \tag{2.36}$$

At low frequencies, the photodiode junction capacitance is negligible and at low detected optical powers there is a negligible voltage drop across the photodiode series resistance. Therefore we are left with the small-signal, low frequency circuit model shown in Fig. 2.16. Again at low frequencies, the frequency dependent components of the load can be neglected, thereby simplifying the photodiode load to a pure resistance R_{LOAD}. The detection power dissipated in this load, p_{load}, is

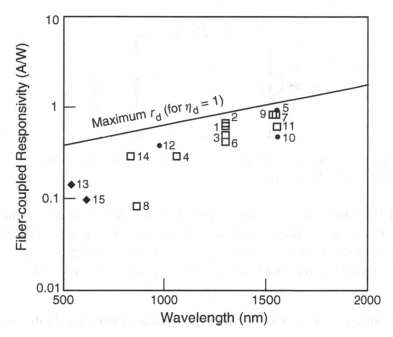

Figure 2.15 Plot of the maximum photodiode responsivity vs. wavelength and points for reported values of several research and commercially available, fiber-coupled photodiodes. ● PIN, □ waveguide, ◆ Schottky/MSM.

Point	Authors and date	Freq. (GHz)	Point	Authors and date	Freq. (GHz)
1	J. Bowers and C. Burrus, 1987	12	7	K. Kato et al., 1991	40
2	Ortel Corporation, 1995	15	8	L. Lin et al., 1996	49
3	A. Williams et al., 1993	20	9	K. Kato et al., 1992	50
4	J. Bowers and C. Burrus, 1986	28	10	D. Wake et al., 1991	50
5	M. Makiuchi et al., 1991	31	11	K. Kato et al., 1994	75
6	J. Bowers and C. Burrus, 1987	36	12	Y. Wey et al., 1993	110
			13	E. Ozbay et al., 1991	150
			14	K. Giboney et al., 1995	172
			15	Y. Chen et al., 1991	375

simply

$$p_{\text{load}} = i_d^2 R_{\text{LOAD}}. \tag{2.37}$$

Squaring (2.36), substituting it into (2.37) for i_d and solving for the ratio $p_{\text{load}}/p_{\text{od}}^2$ yields the expression for the low frequency *incremental detection efficiency* of a photodiode, viz.:

$$\frac{p_{\text{load}}}{p_{\text{od}}^2} = r_d^2 R_{\text{LOAD}}. \tag{2.38}$$

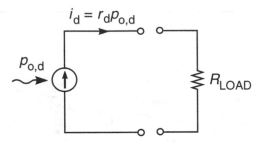

Figure 2.16 Small-signal, low frequency circuit model of a photodiode detector feeding an output load, R_{LOAD}.

The alert reader will notice that in the above derivation we did not include a matching resistor across the photodiode current source. The reason for this involves a more detailed discussion of matching than we have the background to go into at this point. Thus we defer discussion on this point until Section 4.3.1.1.

Appendix 2.1   Steady state (dc) rate equation model for diode lasers

The standard way to relate the external laser variables, such as lasing threshold and optical output power, to the underlying, internal laser parameters is to treat the semiconductor laser as a pair of coupled reservoirs, one for electrons[4] in the active region, and one for photons. The interactions between these two reservoirs can be described mathematically via a pair of coupled differential equations. One differential equation expresses the rate at which carriers are injected and recombine. The other equation of the pair expresses the rate at which photons in the lasing cavity mode are generated and decay. Consequently these equations are known as the *laser rate equations*. In this appendix, we will develop these equations and their steady state solutions. The approach presented here is based on the development of Coldren and Corzine (1995), with the discussion of the laser *P* vs. *I* curve in terms of the steady state solution to the rate equations provided by Lee (1998).

For this development we will assume that there is only one longitudinal optical mode in the cavity, and of course that the cavity is single mode in both the transverse directions as well. Thus this analysis can be applied directly to lasers such as DFBs and VCSELs when lasing in the fundamental mode (see Section 2.2.1). In lasers where a significant fraction of the laser's output is divided among multiple longitudinal modes, these equations may still be applied, but with reservations. For example, in a Fabry–Perot diode laser one can apply these equations to model the

---

[4] Actually there are reservoirs of both electrons and holes. However, because of charge neutrality, the spatial averages of these two types of carriers are equal. Thus we only need to keep track of one type; we chose to refer to the electrons.

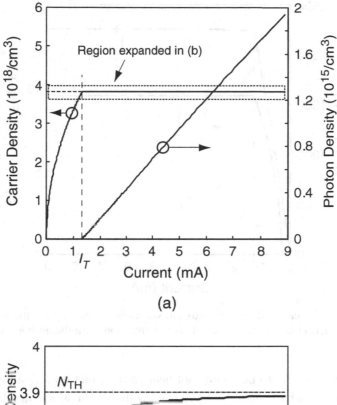

(a)

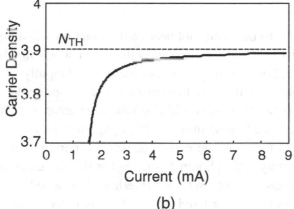

(b)

Figure A2.1.1 (a) Carrier and photon density vs. laser bias current. (b) Expanded carrier density scale showing the "clamping" region. (Lee, 1998, reprinted with permission from the author.)

behavior of the total laser output, i.e. to the sum of the powers in all the lasing modes, and with some restrictions to the individual modes.

Before going through the mathematics, let us discuss the characteristics of the steady state solutions, which are plotted in Fig. A2.1.1.

One of the plots in this figure is the photon density in the mode vs. laser bias current. For this variable we see that below a well defined threshold current the

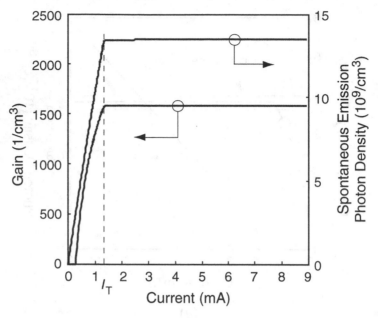

Figure A2.1.2 Gain and spontaneous emission photon density for the same laser as Fig. A2.1.1. (Lee, 1998, reprinted with permission from the author.)

photon density appears to be a constant near zero; in fact it takes a log plot to see that the photon density is actually increasing with current, albeit slowly. Above the threshold current, the photon density increases linearly and rapidly with increasing current. (If one attempted to measure this from a real laser one could get a slightly different result. This is because one would also have the spontaneous emission from the other modes, which we have assumed to be negligible here.)

Also plotted in this figure is the carrier density vs. current. The plot of this variable is complementary to the photon density: below threshold the carrier density increases with laser bias current. But once threshold is reached, we see that the carrier density appears to reach a fixed value, which is independent of laser bias current, i.e. the carrier density appears to be "clamped" at N_{TH}. Because the gain and spontaneous emission are related to the carrier density, these quantities are also "clamped" at their threshold values. This is shown in Fig. A2.1.2. We will use these facts often in later portions of this derivation to simplify our expressions.

It is important to note the relative scales for photon density in Figs. A2.1.1 and A2.1.2. The spontaneous emission, Fig. A2.1.2, is approximately a factor of 10^5 less than the total photon emission, Fig. A2.1.1, which is dominated by stimulated emission.

The basic physical process that causes the steady state solutions to have these characteristics is the competition for injected carriers supplied by the drive current.

At low drive currents, the round trip photon losses exceed the gain and hence the stimulated photons are not able to build and take injected carriers. Thus the injected carriers, which feed the active region carrier reservoir, are removed by the spontaneous emission and non-radiative recombination processes. As the drive current increases, the active region carrier reservoir increases and the optical gain increases. Once the gain nearly equals the loss, the photon density builds very rapidly (exponentially with gain) and the stimulated emission process takes all of the excess injected carriers. That is, once threshold is reached, every electron that enters the active region becomes a photon through stimulated emission. (The dynamics of how threshold is achieved will be discussed in Appendix 4.1.) This accounts for the sharp, linear increase in output power vs. drive current.

It turns out that this also explains why the carrier density is essentially clamped. The key thing to note is that when the carrier density increases by a tiny amount, the photon density – which depends exponentially on the net gain – increases by a lot. Thus, steady state is maintained by a large increase in the photon density and a tiny increase in the carrier density in response to an increase in the drive current. A close examination of Fig. A2.1.1 reveals that the carrier density in fact asymptotically approaches what we have defined as N_{TH}.

We begin the rate equation derivation by examining the change in the carrier density with time, dn_U/dt.[5] This quantity can be expressed as the difference between the rates at which electrons are injected, $r_{U\text{-GEN}}$, and the rate at which these carriers recombine in the active volume of the laser cavity, $r_{U\text{-REC}}$:

$$\frac{dn_U}{dt} = r_{U\text{-GEN}} - r_{U\text{-REC}}. \tag{A2.1.1}$$

The rate at which carriers are injected into the active region simply depends on the fraction, η_i of the total laser current, i_L that flows through the active region divided by the volume, V_E, of that region (assuming that the leakage current is negligible):

$$r_{U\text{-GEN}} = \frac{\eta_i i_L}{q V_E}. \tag{A2.1.2}$$

The multiple mechanisms by which the reservoir of carriers is depleted can be divided into two general categories: radiative and non-radiative. For the moment we will simply denote the rate of carrier recombination via the generation of stimulated emission by r_{ST}, without breaking it down further.

The recombination term, without stimulated emission, can be decomposed into its dominant components: the uni-molecular recombination coefficient, which is proportional to n_U and represents non-radiative recombination through defect states;

---

[5] We use the subscript "U" as the universal reference to carriers, either electrons or holes.

the bi-molecular recombination coefficient, which is proportional to n_U^2 and represents electron–hole recombination events that produce spontaneous emission, r_{SP}; and Auger recombination, which is proportional to n_U^3 and represents non-radiative recombination from collisions between electrons and between holes.

Consequently the non-radiative recombination + spontaneous emission term can be expressed as

$$r_{U\text{-REC-NR}} = An_U + Bn_U^2 + Cn_U^3 = (A + Cn_U^2)n_U + Bn_U^2, \quad (A2.1.3)$$

where the quadratic term represents simply the spontaneous emission rate, r_{SP}.

For any given carrier density, we can associate a characteristic lifetime, $\tau_U(n_U)$ to the term in parentheses in (A2.1.3). Although this greatly simplifies the math, it is important to remember that only in the case of negligible Auger recombination is this lifetime independent of the carrier density.

With this, we can write the total carrier recombination rate as the sum of the rates of the non-radiative, spontaneous and stimulated processes, viz.:

$$r_{U\text{-REC}} = \frac{n_U}{\tau_U} + r_{SP} + r_{ST}. \quad (A2.1.4)$$

Substituting (A2.1.2) and (A2.1.4) into (A2.1.1) yields an expression for the carrier rate equation

$$\frac{dn_U}{dt} = \frac{\eta_i i_L}{q V_E} - \frac{n_U}{\tau_U} - r_{SP} - r_{ST}. \quad (A2.1.5)$$

In an analogous manner we can develop an equation for the rate of change in the number of cavity photons with time, dn_P/dt, as a function of the difference between the rates at which photons are created, $r_{P\text{-GEN}}$, and the rate at which these photons are lost, $r_{P\text{-REC}}$:

$$\frac{dn_P}{dt} = r_{P\text{-GEN}} - r_{P\text{-REC}}. \quad (A2.1.6)$$

Photons are generated via both spontaneous and stimulated recombination processes. Hence our first thought might be that $r_{P\text{-GEN}} = r_{SP} + r_{ST}$. This is almost right; but we need to correct for two factors. One is the fact that the weak optical confinement means that the photon volume, V_P, is typically larger than the electron volume, V_E. Because our development is based on densities – not numbers – of photons, we are interested in the photon density, which is inversely related to V_P, and the action in the cavity, which is inversely related to V_E. Consequently we must define a confinement factor as the ratio of these two volumes, $\Gamma = V_E/V_P \leq 1$, and multiply all terms that are coupled to the carrier rate equation by it.

The other correction factor arises from the fact that the spontaneous emission is omni-directional whereas the stimulated emission is confined to the optical cavity.

Thus we need to define β_{SP}, which is the fraction of the spontaneous emissions that wind up in the lasing mode.

Applying both of these refinements, the complete expression for $r_{P\text{-GEN}}$ is

$$r_{P\text{-GEN}} = \Gamma(\beta_{SP}r_{SP} + r_{ST}). \tag{A2.1.7}$$

Photon losses are dominated by mechanisms such as absorption, scattering and escape from the cavity. Consequently we can characterize these processes collectively by a photon lifetime, τ_P. Thus the expression for $r_{P\text{-GEN}}$ is simply

$$r_{P\text{-REC}} = \frac{n_P}{\tau_P}. \tag{A2.1.8}$$

Of course some of the cavity photons can undergo stimulated absorption to generate carriers. The reason there is no separate rate for this stimulated absorption is that the stimulated emission rate in (A2.1.7) is the difference between the actual stimulated emission and absorption rates.

Substituting (A2.1.7) and (A2.1.8) into (A2.1.6) yields an expression for the photon rate equation

$$\frac{dn_P}{dt} = \Gamma(\beta_{SP}r_{SP} + r_{ST}) - \frac{n_P}{\tau_P}. \tag{A2.1.9}$$

As presented, it appears that (A2.1.5) and (A2.1.9) can be solved to express a number of the laser properties, such as lasing threshold, optical power vs. current, etc., in terms of fundamental material and laser design parameters, such as the carrier densities, cavity mirror reflectivities, etc. However, such is not the case because this initial set of rate equation variables does not explicitly express the two key interdependencies between the number of carriers and the number of photons.

One of the key interdependencies is the fact that the stimulated generation rate depends on the number of photons. Consequently we need an explicit expression for r_{ST} in terms of the number of stimulated photons generated per unit time, viz.:

$$r_{ST} = \frac{\Delta n_P}{\Delta t}. \tag{A2.1.10}$$

Recall that a stimulated photon is an identical copy of the photon that triggered the stimulated photon. Thus the increase in the number of photons, Δn_P, that occurs over an incremental laser cavity length, Δz, can be thought of as the gain, g, that occurs in that length. Expressing this thought in equation form yields

$$n_P + \Delta n_P = n_P e^{g\Delta z}. \tag{A2.1.11}$$

As $\Delta z \to 0$, $\exp(g\Delta z) \to 1 + g\Delta z$. Further, the time, Δt, that it takes the photons to traverse the distance Δz is related by the photon group velocity, v_g, through the simple expression $\Delta z = v_g\Delta t$. Inserting both these relations into (A2.1.11) and solving for the change in the number of stimulated photons per unit time yields the expression we were looking for in (A2.1.10):

$$r_{ST} = \frac{\Delta n_P}{\Delta t} = n_P g v_g. \tag{A2.1.12}$$

The other key interdependence is the fact that the gain depends on the carrier density. The dependence is best approximated analytically by a logarithmic function, which in its most general form is given by

$$g(n_U, N_P) = \frac{g_0}{1 + \varepsilon N_P} \ln \left(\frac{n_U + N_S}{N_T + N_S} \right), \tag{A2.1.13}$$

where N_T is the carrier density at which stimulated emission exactly balances stimulated absorption and the material gain becomes unity, N_S is a fitting parameter that allows modeling of the gain (absorption) when $n_U = 0$ (if we only wish to model the gain for carrier densities above transparency we can set $N_S = 0$ and readjust the other fitting parameters), g_0 is a fitting parameter that models the amount of gain that can be obtained from the material, and ε is the gain compression factor that models the reduction in gain due to the presence of photons. While equation (A2.1.13) models the gain in the semiconductor quite accurately, it is rather complex and can be approximated with much simpler functions. It is common to neglect the gain compression factor and approximate the logarithm by a linear function, viz.:

$$g \cong a(n_U - N_T), \tag{A2.1.13a}$$

where $a = dg/dn_U$ is defined as the differential gain. This approximation is valid because typically the effect of gain compression is rather weak, the carrier density does not exceed the transparency carrier density by very much, and the logarithm is nearly linear in this region. Including these interdependences gives us the following rate equations:

$$\frac{dn_U}{dt} = \frac{\eta_i i_L}{q V_E} - \frac{n_U}{\tau_U} - r_{SP} - v_g g(N_U, N_P) n_P, \tag{A2.1.14a}$$

$$\frac{dn_P}{dt} = \Gamma v_g g(N_U, N_P) n_P + \Gamma \beta_{SP} r_{SP} - \frac{n_P}{\tau_P}. \tag{A2.1.14b}$$

In general the solution to (A2.1.14a and b) is the sum of a steady state and a dynamic solution:

$$n_U = N_U + n_u \tag{A2.1.15a}$$

$$n_P = N_P + n_p. \tag{A2.1.15b}$$

Since this appendix is supporting the static or dc laser response discussion in Chapter 2, we seek only the steady state solution to the rate equations here. We present the dynamic solution in an appendix to Chapter 4.

We are now in a position to derive an expression for the laser P vs. I curve from a steady state solution to the rate equations. It will be most useful to solve for N_P and I_L as the dependent variables with N_U as the independent variable. Setting the time derivatives to zero in (A2.1.14a and b) – since we are seeking the steady state solutions – and applying some simple algebra to (A2.14a and b) leaves us with

$$N_P(N_U) = \frac{\Gamma \beta_{SP} r_{SP}(N_U)}{1/\tau_P - \Gamma v_g g(N_U)} \tag{A2.1.16a}$$

$$I_L(N_U) = \frac{q V_E}{\eta_i} \left(r_{SP} + \frac{N_U}{\tau_U} + v_g g(N_U) N_P(N_U) \right). \tag{A2.1.16b}$$

As expected, these equations reveal the physical processes discussed above. The denominator in the photon density equation, (A2.1.16a), contains the carrier density clamping. It is of course unphysical for the photon density to go to infinity (this would violate the second law of thermodynamics). Thus the threshold gain is defined as

$$g_{TH} = \frac{1}{\Gamma \tau_P v_g}, \tag{A2.1.17}$$

which sets the threshold carrier density N_{TH} through the relationship between gain and carrier density. The asymptotic approach of the carrier density to N_{TH} is evident in equation (A2.1.16a). A physical explanation as to why N_U never equals N_{TH} is that there is always a small fraction of the spontaneous emission that enters the lasing mode, thus the gain never needs to exactly equal the loss for the photons to see a net gain of unity.

Using equation (A2.1.16a) we can rewrite (A2.1.16b) as

$$I_L = \frac{q V_E}{\eta_i} \left(r_{SP} + \frac{N_{TH}}{\tau_U} + v_g g_{TH} N_P \right). \tag{A2.1.18}$$

We now treat I_L as the independent variable and N_P as the dependent variable. We notice that just at threshold, the photon density is zero, which lets us define the threshold current as

$$I_T = \frac{q V_E}{\eta_i} \left(r_{SP} + \frac{N_{TH}}{\tau_U} \right). \tag{A2.1.19}$$

Using this definition, we can rewrite (A2.1.16a) as

$$N_P = \frac{\eta_i}{q V_E v_g g_{TH}} (I - I_T) = \frac{\tau_P \eta_i}{q V_P} (I - I_T), \tag{A2.1.20}$$

which reveals a linear relationship between the photon density and the drive current above threshold. This is clearly evident in the plot of Fig. A2.1.1.

Thus far, all of our expressions involve the internal photon density, which is not a quantity we can readily measure. Output optical power is a readily measurable quantity that can be easily related to photon density via the appropriate unit conversion. Since power is in J/s, we note the following:

$$
P_O \left[\frac{J}{s} \right] = \left\{ N_P \left[\frac{photons}{cm^3} \right] \right\} \left\{ h\nu \left[\frac{J}{photon} \right] \right\} \left\{ V_P \, [cm^3] \right\} \left\{ v_g \alpha_m \left[\frac{1}{s} \right] \right\} F,
$$

(A2.1.21)

where the first three terms on the right hand side are the energies in the cavity and the fourth term is the rate at which the energy leaves the cavity. The factor F represents the fraction of the optical power that leaves the facet where the light is measured. For a cavity with equal reflectivities at both ends, $F = 1/2$.

Substituting the second form of (A2.1.20) into (A2.1.21) yields

$$
P_O = h\nu v_g \alpha_m F \left(\frac{\tau_P \eta_i}{q} \right) (I - I_T).
$$

(A2.1.22)

This is close to the final result we are seeking. We simply need to eliminate the explicit dependency of (A2.1.22) on photon lifetime that we previously defined in terms of the decay of photons out of the cavity. We do this by relating τ_p to the optical losses of the cavity: the internal material scattering loss, α_i (1/cm), and the mirror loss, α_m (1/cm). The total cavity loss is $(\alpha_i + \alpha_m)$ (1/cm), which when converted to a loss rate through the group velocity, we recognize as the photon lifetime:

$$
\tau_P = \frac{1}{v_g(\alpha_i + \alpha_m)}.
$$

(A2.1.23)

By substituting (A2.1.23) into (A2.1.22), we obtain an expression for the output optical power as a function of the bias current:

$$
P_O = \eta_i \frac{h\nu}{q} \frac{\alpha_m}{\alpha_i + \alpha_m} (I - I_T).
$$

(A2.1.24)

The term $\alpha_m/(\alpha_i + \alpha_m)$ is the external quantum efficiency and is often written η_d. Physically, it represents the fraction of the photons that escape through the mirrors.

The mirror loss can be related to the mirror reflectivities at each end of the cavity, R_1 and R_2. Alternatively, the mirror loss can be viewed as simply the discrete mirror loss at the ends of the cavity distributed throughout the cavity length. Since these are two views of the same process, we can arrive at an expression for α_m in terms of R_1 and R_2 by equating the distributed round trip loss to the discrete round trip loss:

$$e^{\alpha_m 2L} = \frac{1}{R_1 R_2} \Rightarrow \alpha_m = \frac{1}{2L} \ln\left(\frac{1}{R_1 R_2}\right). \qquad (A2.1.25)$$

If the mirror reflectivity is high, α_m is small and the external efficiency is low. This is because the photons are trapped inside the cavity longer and have more opportunity to be lost through internal scattering.

Appendix 2.2 Absorption coefficient of an electro-absorption modulator

The absorption as a function of wavelength, or equivalently frequency, of the incident optical wave is a complex function of the separation between the incident photon's frequency, ω, the material bandgap frequency, ω_g, and the frequency shift caused by the applied electric field, ω_E (Welstand, 1997):

$$\alpha(v_M) \propto \left(\frac{\omega - \omega_g}{\omega_E}|Ai(\beta)|^2 + \sqrt{\omega_E}|Ai'(\beta)|^2\right), \qquad (A.2.2.1)$$

where $Ai(x)$ is the Airy function (see for example Abramowitz and Stegun, 1964),

$$\beta = \hbar\frac{\omega_g - \omega}{\omega_E}, \qquad (A.2.2.2)$$

and

$$\omega_E = \sqrt[3]{\frac{e^2 E^2}{2\mu\hbar}}. \qquad (A.2.2.3)$$

The electric field, E, across the absorbing region can be approximated as the applied modulation voltage, v_M, divided by the width of the intrinsic layer, d; i.e.

$$E \cong \frac{v_M}{d}. \qquad (A.2.2.4)$$

References

Abramowitz, M. and Stegun, I. A. 1964. *Handbook of Mathematical Functions*, New York: Dover Publications, Section 10.4, p. 446.

Ackerman, E. I. 1997. Personal communication.

Agrawal, G. P. and Dutta, N. K. 1986. *Long-Wavelength Semiconductor Lasers*, New York: Van Nostrand Reinhold.

Alferness, R. C. 1982. Waveguide electrooptic modulators, *IEEE Trans. Microwave Theory Tech.*, **30**, 1121–37.

Betts, G. E., Johnson, L. M. and Cox, C. H. III 1988. High-sensitivity bandpass RF modulator in LiNbO$_3$, *Proc. SPIE*, **993**, 110–16.

Bridges, W. B. and Schaffner, J. H. 1995. Distortion in linearized electrooptic modulators, *IEEE Trans. Microwave Theory Tech.*, **43**, 2184–97.

Chen, T. R., Eng, L. E., Zhao, B., Zhuanag, Y. H. and Yariv, A. 1993. Strained single quantum well InGaAs lasers with a threshold current of 0.25 mA, *Appl. Phys. Lett.*, **63**, 2621–3.

Choquette, K. and Hou, H. 1997. Vertical-cavity surface emitting lasers: moving from research to manufacturing, *Proc. IEEE*, **85**, 1730–9.

Coldren, L. A. and Corzine, S. W. 1995. *Diode Lasers and Photonic Integrated Circuits*, New York: John Wiley & Sons.

Cox, C. H., III, 1996. Optical transmitters. In *The Electronics Handbook*, J. C. Whitaker, ed., Boca Raton, FL: CRC Press, Chapter 57.

Cox, C. H., III, Ackerman, E. I. and Betts, G. E. 1996. Relationship between gain and noise figure of an optical analog link, *IEEE MTT-S Symp. Dig.*, 1551–4.

Gradshteyn I. S. and Ryzhik I. M. 1965. *Tables of Integrals, Series and Products*, 4th edition, New York: Academic Press, equation 1.412.1.

Graham, C. H., Bartlett, N. R., Brown, J. L., Hsia, Y., Mueller, C. G. and Riggs, L. A. 1965. *Vision and Visual Perception*, New York: John Wiley & Sons, 351–3.

Gray, P. E. and Searle, C. L. 1969. *Electronic Principles: Physics, Models and Circuits*, New York: John Wiley & Sons, Section 11.4.1.

Halemane, T. R. and Korotky, S. K. 1990. Distortion characteristics of optical directional coupler modulators, *IEEE Trans. Microwave Theory Tech.*, **38**, 669–73.

IEEE 1964. IEEE standard letter symbols for semiconductor devices, *IEEE Trans. Electron Devices*, **11**, no. 8.

Knupfer, B., Kiesel, P., Kneissl, M., Dankowski, S., Linder, N., Weimann, G. and Dohler, G. H. 1993. Polarization-insensitive high-contrast GaAs/AlGaAs waveguide modulator based on the Franz-Keldysh effect, *IEEE Photon. Technol. Lett.*, **5**, 1386–8.

Lee, H. 1998. Personal communication.

Martin, W. E. 1975. A new waveguide switch/modulator for integrated optics, *Appl. Phys. Lett.*, **26**, 562–3.

Papuchon, M., Roy, A. M. and Ostrowsky, B. 1977. Electrically active optical bifurcation: BOA, *Appl. Phys. Lett.*, **31**, 266–7.

Thompson, G. H. B. 1980. *Physics of Semiconductor Laser Devices*, New York: John Wiley & Sons.

Welstand, R. 1997. High linearity modulation and detection in semiconductor electroabsorption waveguides, Ph.D. dissertation, University of California, San Diego, Chapter 3, pp. 62–4.

Yang, G. M., MacDougal, M. H. and Dapkus, P. D. 1995. Ultralow threshold current vertical-cavity surface-emitting lasers obtained with selective oxidation, *Electron. Lett.*, **31**, 886–8.

Yu, P. K. L. 1997. Optical receivers. In *The Electronics Handbook*, Florida: CRC Press, Chapter 58.

References for Fig. 2.3: Laser efficiency vs. wavelength and frequency (Fig. 4.2)

1. M. Peters, M. Majewski and L. Coldren, Intensity modulation bandwidth limitations of vertical-cavity surface-emitting laser diodes, *Proc. IEEE LEOS Summer Topical Meeting (LEOS-STM'93)*, March 1993, pp. 111–13.
2. Fujitsu, Fujitsu Laser Model FLD3F7CX, 1996.
3. B. Moller, E. Zeeb, T. Hackbarth and K. Ebeling, High speed performance of 2-D vertical-cavity laser diode arrays, *IEEE Photon. Technol. Lett.*, **6** (1994), 1056–8.

4. T. Chen, Y. Zhuang, A. Yariv, H. Blauvelt and N. Bar-Chaim. Combined high power and high frequency operation of InGaAsP/InP lasers at 1.3 μm, *Electron. Lett.*, **26** (1990), 985–7.

5. Y. Nakano, M. Majewski, L. Coldren, H. Cao, K. Tada and H. Hosomatsu. Intrinsic modulation response of a gain-coupled MQW DFB laser with an absorptive grating, *Proc. Integrated Photonics Research Conf.*, March 1993, pp. 23–6.

6. W. Cheng, K. Buehring, R. Huang, A. Appelbaum, D. Renner and C. Su. The effect of active layer doping on static and dynamic performance of 1.3 μm InGaAsP lasers with semi-insulating current blocking layers, *Proc. SPIE*, **1219** (1990).

7. T. Chen, P. Chen, J. Ungar and N. Bar-Chaim. High speed complex-coupled DFB laser at 1.3 μm, *Electron. Lett.*, **30** (1994), 1055–7.

8. H. Lipsanen, D. Coblentz, R. Logan, R. Yadvish, P. Moreton and H. Temkin. High-speed InGaAsP/InP multiple-quantum-well laser, *IEEE Photon. Technol. Lett.*, **4** (1992), 673–5.

9. T. Chen, J. Ungar, X. Yeh and N. Bar-Chaim. Very large bandwidth strained MQW DFB laser at 1.3 μm, *IEEE Photon. Technol. Lett.*, **7** (1995), 458–60.

10. 10. R. Huang, D. Wolf, W. Cheng, C. Jiang, R. Agarwal, D. Renner, A. Mar and J. Bowers. High-speed, low-threshold InGaAsP semi-insulating buried crescent lasers with 22 GHz bandwidth, *IEEE Photon. Technol. Lett.*, **4** (1992), 293–5.

11. L. Lester, S. O'Keefe, W. Schaff and L. Eastman. Multiquantum well strained layer lasers with improved low frequency response and very low damping, *Electron. Lett.*, **28** (1991), 383–5.

12. Y. Matsui, H. Murai, S. Arahira, S. Kutsuzawa and Y. Ogawa. 30-GHz bandwidth 1.55 μm strain-compensated InGaAlAs-InGaAsP MQW laser, *IEEE Photon. Technol. Lett.*, **9** (1997), 25–7.

13. J. Ralston, S. Weisser, K. Eisele, R. Sah, E. Larkins, J. Rosenzweig, J. Fleissner and K. Bender. Low-bias-current direct modulation up to 33 GHz in InGaAs/GaAs/AlGaAs pseudomorphic MQW ridge-waveguide devices, *IEEE Photon. Technol. Lett.*, **6** (1994), 1076–9.

14. S. Weisser, E. Larkis, K. Czotscher, W. Benz, J. Daleiden, I. Esquivias, J. Fleissner, J. Ralston, B. Romero, R. Sah, A. Schonfelder and J. Rosenzweig. Damping-limited modulation bandwidths up to 40 GHz in undoped short-cavity multiple-quantum-well lasers, *IEEE Photon. Technol. Lett.*, **8** (1996), 608–10.

References for Fig. 2.10: Modulator efficiency vs. wavelength and frequency (Fig. 4.6)

1. C. Gee, G. Thurmond and H. Yen. 17-GHz bandwidth electro-optic modulator, *Appl. Phys. Lett.*, **43** (1993), 998–1000.

2. C. Cox, E. Ackerman and G. Betts. Relationship between gain and noise figure of an optical analog link, *IEEE MTT-S Digest* (1996), 1551–4.

3. G. Betts, L. Johnson and C. Cox. High-sensitivity lumped-element bandpass modulators in LiNbO₃, *IEEE J. Lightwave Technol.* (1989), 2078–83.

4. R. Jungerman and D. Dolfi. Lithium niobate traveling-wave optical modulators to 50 GHz, *Proc. IEEE LEOS Summer Topical Meeting (LEOS-STM'92)*, August (1992), 27–8.

5. A. Wey, J. Bristow, S. Sriram and D. Ott. Electrode optimization of high speed Mach–Zender interferometer, *Proc. SPIE*, **835** (1987), 238–45.

6. K. Kawano, T. Kitoh, H. Jumonji, T. Nozawa, M Yanagibashi and T. Suzuki. Spectral-domain analysis of coplanar waveguide traveling-wave electrodes and their applications to Ti:LiNbO$_3$ Mach–Zehnder optical modulators, *IEEE Trans. Microwave Theory Tech.*, **39** (1991), 1595–601.

7. R. Madabhushi. Wide-band Ti: LiNbO$_3$ optical modulator with low driving voltage, *Proc. Optical Fiber Communications Conf. (OFC '96)*, 206–7.

8. K. Noguchi, O. Mitomi, K. Kawano and M. Yanagibashi. Highly efficient 40-GHz bandwidth Ti:LiNbO$_3$ optical modulator employing ridge structure, *IEEE Photon. Technol. Lett.*, **5** (1993), 52–4.

9. K. Kawano, T. Kitoh, H. Jumonji, T. Nozawa and M. Yanagibashi. New travelling-wave electrode Mach–Zehnder optical modulator with 20 GHz bandwidth and 4.7 V driving voltage at 1.52 μm wavelength, *Electron. Lett.*, **25** (1989), 1382–3.

10. M. Rangaraj, T. Hosoi and M. Kondo. A wide-band Ti:LiNbO$_3$ optical modulator with a conventional coplanar waveguide type electrode, *IEEE Photon. Technol. Lett.*, **4** (1992), 1020–2.

11. K. Noguchi, K. Kawano, T. Nozawa and T. Suzuki. A Ti:LiNbO$_3$ optical intensity modulator with more than 20 GHz bandwidth and 5.2 V driving voltage, *IEEE Photon. Technol. Lett.*, **3** (1991), 333–5.

12. K. Noguchi, H. Miyazawa and O. Mitomi. 75 GHz broadband Ti:LiNbO$_3$ optical modulator with ridge structure, *Electron. Lett.*, **30** (1994), 949–51.

13. O. Mikami, J. Noda and M. Fukuma (NTT, Musashino, Japan). Directional coupler type light modulator using LiNbO$_3$ waveguides, *Trans. IECE Japan*, **E-61** (1978), 144–7.

14. C. Rolland, G. Mak, K. Prosyk, C. Maritan and N. Puetz. High speed and low loss, bulk electroabsorption waveguide modulators at 1.3 μm, *IEEE Photon. Technol. Lett.*, **3** (1991), 894–6.

15. Y. Liu, J. Chen, S. Pappert, R. Orazi, A. Williams, A. Kellner, X. Jiang and P. Yu. Semiconductor electroabsorption waveguide modulator for shipboard analog link applications, *Proc. SPIE*, **2155** (1994), 98–106.

16. T. Ido, H. Sano, D. J. Moss, S. Tanaka and A. Takai. Strained InGaAs/InAlAs MQW electroabsorption modulators with large bandwidth and low driving voltage, *IEEE Photon. Technol. Lett.*, **6** (1994), 1207–9.

17. I. Kotaka, K. Wakita, K. Kawano, M. Asai and M. Naganuma. High speed and low-driving voltage InGaAs/InAlAs multiquantum well optical modulators, *Electron. Lett.*, **27** (1991), 2162–3.

18. T. Ido, H. Sano, M. Suzuki, S Tanaka and H. Inoue. High-speed MQW electroabsorption optical modulators integrated with low-loss waveguides, *IEEE Photon. Technol. Lett.*, **7** (1995), 170–2.

19. F. Devaux, P. Bordes, A. Ougazzaden, M. Carre and F. Huet. Experimental optimization of MQW electroabsorption modulators with up to 40 GHz bandwidths, *Electron. Lett.*, **30** (1994), 1347–8.

References for Fig. 2.15: Detector efficiency vs. wavelength and frequency (Fig. 4.8)

1. J. Bowers and C. Burrus. Ultrawide-band long-wavelength p-i-n photodetectors, *J. Lightwave Technol.*, **15** (1987), 1339–50.

2. Ortel Corporation. Microwave FP Laser Transmitters, 1530B, *Microwaves on Fibers Catalog*, 1995.

3. A. Williams, A. Kellner and P. Yu. High frequency saturation measurements of an InGaAs/InP waveguide photodetector, *Electron. Lett.*, **29** (1993), 1298–9.
4. J. Bowers and C. Burrus. Heterojunction waveguide photodetectors, *Proc. SPIE*, **716** (1986), 109–13.
5. M. Makiuchi, H. Hamaguchi, T. Mikawa and O. Wada. Easily manufactured high-speed back-illuminated GaInAs/InP p-i-n photodiode, *IEEE Photon. Technol. Lett.*, **3** (1991), 530–1.
6. J. Bowers and C. Burrus. Ultrawide-band long-wavelength p-i-n photodetectors, *J. Lightwave Technol.*, **15** (1987), 1339–50.
7. K. Kato, S. Hata, A. Kozen, J. Yoshida and K. Kawano. High-efficiency waveguide InGaAs pin photodiode with bandwidth of over 40 GHz, *IEEE Photon. Technol. Lett.*, **3** (1991), 473–5.
8. L. Lin, M. Wu, T. Itoh, T. Vang, R. Muller, D. Sivco and A. Cho. Velocity-matched distributed photodetectors with high-saturation power and large bandwidth, *IEEE Photon. Technol. Lett.*, **8** (1996), 1376–8.
9. K. Kato, S. Hata, K. Kawano, H. Yoshida and A. Kozen. A high-efficiency 50 GHz InGaAs multimode waveguide photodetector, *IEEE J. Quantum Electron.*, **28** (1992), 2728–35.
10. D. Wake, T. Spooner, S. Perrin and I. Henning. 50 GHz InGaAs edge-coupled pin photodetector, *Electron. Lett.*, **27** (1991), 1073–5.
11. K. Kato, A. Kozen, Y. Maramoto, T. Nagatsuma and M. Yaita. 110-GHz, 50%-efficiency mushroom-mesa waveguide p-i-n photodiode for a 1.55-μm wavelength, *IEEE Photon. Technol. Lett.*, **6** (1994), 719–21.
12. Y. Wey, K. Giboney, J. Bowers, M. Rodwell, P. Silvestre, P. Thiagarajan and G. Robinson. 108-GHz GaInAs/InP p-i-n photodiodes with integrated bias tees and matched resistors, *IEEE Photon. Technol. Lett.*, **5** (1993), 1310–12.
13. E. Ozbay, K. Li and D. Bloom. 2.0 ps, 150 GHz GaAs monolithic photodiode and all-electronic sampler, *IEEE Photon. Technol. Lett.*, **3** (1991), 570–2.
14. K. Giboney, R. Nagarajan, T. Reynolds, S. Allen, R. Mirin, M. Rodwell and J. Bowers. Travelling-wave photodetectors with 172-GHz bandwidth-efficiency product, *IEEE Photon. Technol. Lett.*, **7** (1995), 412–14.
15. Y. Chen, S. Williamson, T. Brock, R. Smith and A. Calawa. 375-GHz-bandwidth photoconductive detector, *Appl. Phys. Lett.*, **59** (1991), 1984–6.

3

Low frequency, short length link models

3.1 Introduction

The devices discussed in Chapter 2 are rarely used individually. More commonly a modulation device – either a diode laser or an external modulator – is combined with a photodetection device to form a link. In this chapter we begin to examine the performance of complete links by developing expressions for the gain of a link in terms of the modulation and photodetection device parameters. In subsequent chapters we develop analogous expressions for link frequency response, noise figure and dynamic range.

Recall from Chapter 1 that we defined a link as comprising all the components necessary to convey an electrical signal over an optical carrier. Since the definition of available power requires an impedance match, we expand the link definition slightly to include those passive electrical components needed to impedance match the modulation and photodetection devices to the electrical signal source and load, respectively. The impedance matching function is also required by the definitions of some of the link parameters we will be discussing. A more detailed version of the link block diagram is shown in Fig. 3.1.

Although the models we develop have applicability at any frequency, we choose to focus on relatively low frequencies here where lumped-element RLC passive elements are appropriate. This permits us to get the important concepts across without their being obscured by the myriad detailed effects that microwave models require. For similar reasons we limit the scope of our discussion at this point to lengths of fiber where their various deleterious effects can be neglected as well.

The emphasis in this and subsequent chapters is on the response of the link to small-signals. For many cases the intuitive notion of what constitutes a "small-signal" is sufficient. However, there are clearly cases when it is important to be able to define the range of link inputs over which the small-signal approximation is

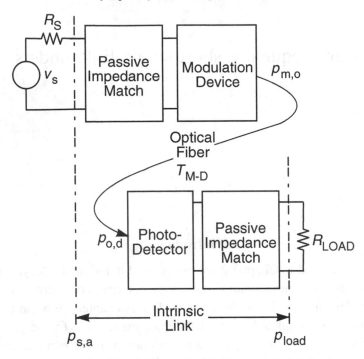

Figure 3.1 Block diagram of the basic optical link.

valid. Hence we also develop large-signal intrinsic gain models in the last section of this chapter.

In general, gain, from a mathematical perspective, is a complex quantity, i.e. it has both a magnitude and a phase. However, in this chapter we will consider only the magnitude of the gain; the complex gain will be discussed in the next chapter. Consistent with our assumption to neglect the optical fiber loss between modulation and photodetection devices, the gain expressions do not include the effects of phase shift introduced by the fiber.

3.2 Small-signal intrinsic gain

A basic link parameter is the link gain, which derives its importance not only from its own merits but also from the fact that several other link parameters depend upon it. In the RF field there are numerous definitions of gain (see, for example, Linvill and Gibbons, 1961 or Pettai, 1984). The one that is most applicable to analog links is the *transducer power gain*, g_t. This is defined as the ratio of the power delivered into a matched load, p_{load} to the available power of the source, $p_{s,a}$:

$$g_t \equiv \frac{p_{load}}{p_{s,a}}. \tag{3.1}$$

Since the transducer gain is the one we refer to most often, we omit "transducer" and simply refer to this quantity as gain. There is no requirement from the definition that the gain be greater than one. Consequently we will often use gain in the general sense where gains less than 1 will represent losses. There often is need to express gains in decibels,

$$G_t = 10 \log_{10}(g_t), \tag{3.2}$$

where we have used G_t to denote the gain g_t in dB. Since log (1) $= 0$, a lossless link, $g_t = 1$, has 0 dB gain.

Gains greater than one can be achieved in many ways. Since our focus here is on understanding the impacts of the modulation and photodetection devices on the link gain, we first study the link gain without the obscuring effects of amplifiers. Consequently we define the *intrinsic link gain*, g_i, to be the transducer power gain of an amplifierless optical link. Since impedance matching is necessary to the definition of transducer gain, we include passive impedance matching in the definition of intrinsic gain. The tradeoffs between link and amplifier parameters will be discussed in Chapter 7.

To form a link we simply need to couple the output of the modulation device to the input of the photodetection device. In general there are some losses in this process. We represent the net effect of these optical losses by the optical transmission, T_{M-D}, between the modulation and photodetection devices. Clearly without any optical amplifiers $0 \leq T_{M-D} \leq 1$. We can use this optical transmission to couple the optical output of the modulation device to the optical input of the photodetection device:

$$p_{o,d} = T_{M-D} p_{m,o}. \tag{3.3}$$

We can now use (3.3), together with the material developed in Chapter 2, to express (3.1) as the product of the incremental modulation efficiency for the modulator and the incremental detection efficiency for the photodetector, viz.:

$$g_i = \left(\frac{p_{m,o}^2}{p_{s,a}} \right) T_{M-D}^2 \left(\frac{p_{load}}{p_{d,o}^2} \right). \tag{3.4}$$

This equation should reassure those who were concerned that the squared terms that appeared in the individual incremental modulation and detection efficiency expressions would lead to a link with a non-linear transfer function between electrical input and output. As (3.4) shows, the modulation device does establish a square dependence between electrical and optical power, but the photodetector reverses the modulation function by having the exact inverse square relationship between optical and electrical powers. Consequently these links are, to first order, linear. In Chapter 6 we will explore distortion in analog links, but this distortion will be caused

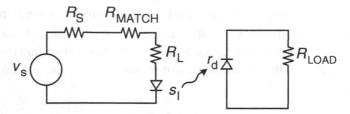

Figure 3.2 Schematic diagram of a direct modulation link based on a diode laser with resistive matching and a photodiode detector.

by non-linearities in the device transfer functions, not the quadratic dependences of (3.4).

Expressing (3.4) in terms of dB brings another important point to light:

$$G_i = 10 \log \left(\frac{p_{m,o}^2}{p_{s,a}} \right) + 20 \log(T_{M-D}) + 10 \log \left(\frac{p_{load}}{p_{d,o}^2} \right). \tag{3.5}$$

Because of the square of optical transmission in (3.4), (3.5) makes clear that a 1 dB change in this transmission results in a 2 dB change in the link intrinsic gain. For example in a distribution application such as CATV, the 12 dB optical loss of an ideal 1:16 optical splitter contributes 24 dB of loss to the intrinsic gain.

To avoid distracting the focus of the development in this chapter, we assume that $T_{M-D} = 1$ in the following.

3.2.1 Direct modulation

A direct modulation link consists of a diode laser and a photodetection device, such as the ones discussed in Chapter 2. For specificity in the discussion below, we choose a PIN photodiode for the photodetection device here, since this is the predominant choice in present link designs. These devices, along with the impedance matching resistor for the diode laser, form a link as shown in Fig. 3.2.

Expressions for both the ratios in (3.4) were derived in Chapter 2; the first ratio in (2.11), the second ratio in (2.38). Substituting these equations into (3.4) yields

$$g_i = \left(\frac{s_\ell^2}{R_L + R_{MATCH}} \right) \left(r_d^2 R_{LOAD} \right). \tag{3.6}$$

An important special case that is often encountered in practice occurs when $R_L + R_{MATCH} = R_{LOAD}$. Applying this constraint to (3.6) reduces the equation for link gain to simply the product of the square of the slope efficiency and responsivity, viz.:

$$g_i = s_\ell^2 r_d^2. \tag{3.7}$$

The fact that the slope efficiency and responsivity appear as the square is a consequence of the fact that the modulated optical power from the diode laser is proportional to the electrical current through it and vice versa for the photodetector.

To get a feel for the range of intrinsic gains that are achievable at low frequencies with present diode lasers and photodiodes, we can refer to the values of diode laser slope efficiency and photodetector responsivity shown in Figs. 2.3 and 2.15, respectively. For example, at an optical wavelength of 1.3 μm, the diode laser fiber-coupled slope efficiencies range from 0.035 to 0.32 W/A while the photodiode responsivities range from 0.5 to 0.8 A/W. Using (3.5) we see that these values of slope efficiency and responsivity correspond to intrinsic link gains between -35 and -12 dB.

These losses, while substantial, can be overcome with commercially available electronic amplifiers. If one is solely interested in overcoming the loss, it makes no difference whether pre- or post-link amplifiers are used. However, other link parameters, such as noise figure, are highly dependent on whether pre- or post-amplification is used. Hence we defer a more complete discussion of these tradeoffs to Section 7.3.

Alternatively we could consider more efficient means of impedance matching, both at the laser and the photodiode to overcome the intrinsic link loss. Since such approaches invariably involve a bandwidth tradeoff, we defer discussion of them until Chapter 4. A fundamentally more satisfying approach would be to improve the device fiber-coupled slope efficiencies, especially of the diode laser or the diode laser slope efficiency as has been recently reported by Cox *et al.* (1998).

Another approach to increase the intrinsic gain is to use a different impedance matching approach. For example, the link loss could be reduced if we replace the lossy resistive match, R_{MATCH}, with a lossless transformer. Since this approach affects the bandwidth, we defer discussion of it until we have developed the frequency response of a link in Chapter 4.

It is tempting to increase the intrinsic gain by setting $R_{MATCH} = 0$ without replacing it with any other form of impedance matching. This does indeed increase the intrinsic gain – by a maximum of 6 dB when R_L also equals zero – since the current through the laser can at most double. However, the increased gain is achieved with the penalty of an increased percentage of the incident RF power reflected back into the modulation source. If the modulation source can tolerate this mismatch, then one could consider this approach. Alternatively one could consider the design of a modulation source with internal impedance R_L. More often in designing a link we need to include consideration of noise and distortion in addition to link gain. In these cases it is generally better to maintain an impedance match between source and modulation device.

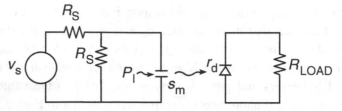

Figure 3.3 Schematic diagram of external modulation link using a Mach–Zehnder modulator with resistive match and a photodiode detector.

3.2.2 External modulation

An external modulation link consists of a CW laser, a modulator and a photodetection device, such as the ones discussed in Chapter 2. For specificity in the discussion below, we will choose a Mach–Zehnder modulator (MZM) with lumped-element electrodes for the modulator and a PIN photodiode for the photodetection device here, since these are the predominant choices in present external modulation link designs. These devices, along with the impedance matching resistor for the MZM form a link as shown in Fig. 3.3.

To obtain an explicit expression for link gain, we begin by substituting into (3.4) the expressions from Chapter 2 for the incremental modulation and detection efficiencies in terms of the slope efficiency for the MZM, (2.17), and responsivity for the photodiode, (2.38). The result is the following:

$$g_i = \left(\frac{s_m^2}{R_S}\right)\left(r_d^2 R_{LOAD}\right). \tag{3.8}$$

If we again assume that the source and load impedances are equal, we see that they once again cancel out, leaving the expression for external modulation link gain dependent only on the square of the modulation slope efficiency and photodiode responsivity:

$$g_i = s_m^2 r_d^2. \tag{3.9}$$

Thus we reach the important conclusion that the expression for intrinsic gain in terms of slope efficiency and responsivity is the same for *any* intensity modulated link; direct or external, Mach–Zehnder, directional coupler or electro-absorption modulator.

Although (3.9) has the same form as (3.7), recall from Chapter 2 that the external modulation slope efficiency is not constrained to be less than or equal to one, even in the impedance matched case, as was the case for the slope efficiency for direct modulation using conventional lasers. Thus an external modulation link has the potential for intrinsic gain greater than one, i.e. $g_i > 1$, even when using a PIN photodiode. This conclusion is not dependent on the specifics of a MZM, but is true for all known types of external modulators.

To get a feel for the range of external modulation intrinsic gains that are achievable today, we can use the slope efficiency and responsivity presented in Figs. 2.10 and 2.15 for the modulator and photodiode, respectively. To make these results comparable to the direct modulation gains presented above, we again assume that we are working at an optical wavelength of 1.3 μm. The external modulation slope efficiencies range from 0.06 to 70 W/A, which results in a range of external modulation intrinsic gains from −30 dB to +35 dB. Thus, in contradistinction to the direct modulation gains, we see that the higher modulator slope efficiencies can result in substantially lower intrinsic link loss or even significant gain.

Of course one could also use any of the techniques discussed in Section 3.2.1 for improving the intrinsic gain of a direct modulation link – i.e. electronic pre- or post-amplification either separately or together with lossless impedance matching – to increase the gain of an external modulation link as well.

3.3 Scaling of intrinsic gain

We have shown via (3.7) and (3.9) that it is possible to use the same functional form to express the intrinsic gain for both direct and external modulation links. This form is useful for some aspects of link design, particularly in cases where the emphasis is on the slope efficiency *per se* and not how a particular value of slope efficiency is achieved. As an example of the utility of this formalism, we examine the scaling of link gain with changes in modulation slope efficiency and photodetection responsivity.

However, the above formalism does mask some important insights that are only revealed when we examine the role various device parameters play in determining the individual slope efficiency and responsivity. Consequently there are times in link design when it is useful, or even essential, to exploit these underlying relationships to meet a link design goal. We give two examples of this approach by examining the scaling of link gain with optical power and wavelength.

Which of these two general approaches to use in any given link design situation is, of course, up to the link designer. Maturity in link design, as with other areas of analysis and design, is achieved when the designer knows which tools, from among those in the link design tool box, are the best to use in any given situation.

3.3.1 Optical power

One of the key reasons for the difference in gains between direct and external modulation links can be traced to the different dependence on optical power between

Table 3.1 *Modulation device slope efficiencies and detector responsivity for intrinsic gain curves plotted in Fig. 3.4*

Parameter		Value
Direct modulation	s_{l-FP}	0.05 W/A
	s_{l-DFB}	0.3 W/A
	r_d	1.0 A/W
External modulation	s_{mz}	2.4 W/A at $I_D = 15.5$ mA
	r_d	0.85 A/W

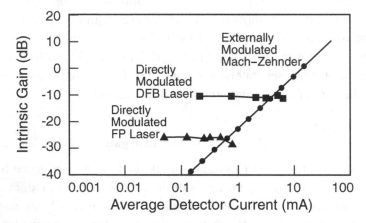

Figure 3.4 Plot of intrinsic link gain vs. average optical power – as measured by the average photodetector current – for direct modulation with Fabry–Perot (FP) and distributed feedback (DFB) lasers; and external modulation using a Mach–Zehnder modulator. Experimentally measured values of intrinsic link gain shown by symbols.

the two types of links. To explore the scaling of direct and external modulation link gain with average optical power, we plot intrinsic gain vs. average optical power for both types of links. As we have throughout this chapter, we will assume a PIN photodetector and resistive impedance matching to the laser or external modulator. Recall from (2.36) that the photodiode responsivity does not depend on optical power over at least several orders of magnitude. Thus any optical power scaling effects we see are due to the modulation device.

For a direct modulation link we select representative values of diode laser slope efficiency and photodetector responsivity from the data shown in Figs. 2.3 and 2.15, respectively. The values selected for the present calculations are listed in Table 3.1, and the link gains calculated from substituting them into (3.7) are plotted by the lines in Fig. 3.4; experimentally measured values of intrinsic link gain are also plotted in Fig. 3.4 by the symbols.

Recall from (2.5) that s_ℓ is approximately independent of optical power from the lasing threshold until it begins to decrease at high bias currents. Consequently the intrinsic gain is at best independent of average optical power. In practice this gain decreases following the decrease in slope efficiency at high optical power, as shown by the Fabry–Perot laser data in Fig. 3.4.

To determine the scaling of external modulation intrinsic gain with optical power, we will substitute the explicit expression for MZM slope efficiency, (2.18), into (3.9); the result is

$$g_{\mathrm{mzpd}} = \left(\frac{T_{\mathrm{FF}} P_{\mathrm{I}} \pi R_{\mathrm{S}}}{2 V_\pi} \right)^2 r_{\mathrm{d}}^2.$$

(3.10)

The external modulation intrinsic gain increases as the square of average optical power as indicated by (3.10). Using the representative set of values listed in Table 3.1, we also plot in Fig. 3.4 the intrinsic gain for this external modulation link. As expected, the external modulation link gain increases as the square of average optical power. What may not initially be expected is that at sufficiently high optical powers the intrinsic gain is greater than one, indicating net power gain from the link.

Achieving positive power gain from what may at first appear to be a collection of passive components can be troubling, so we present a plausibility justification for the gain. Recall from (3.1) that gain is defined as the ratio of the power delivered to a load to the available source power. At the link input, the modulator impedance is determined only by the substrate parameters and the geometry of the modulator electrodes. This impedance is therefore independent of the level of optical power flowing through the modulator waveguides. Consequently the available source power drawn by the modulator is also independent of the optical power flowing through the modulator. Conversely, at the output end of the link, the modulated power delivered to the photodetector load is clearly dependent on the level of optical power incident on the detector. In the extreme case of no optical power on the detector, no RF power is delivered to the load. Thus as one increases the average optical power through the modulator, the detected output power increases while the link input power remains constant. Eventually the detected output power exceeds the modulator input power, resulting in net modulated power gain from the link.

Viewed from this perspective, obtaining gain from a link is no more unusual than obtaining gain from a transistor. Figure 3.5(a) shows a field effect transistor (FET) connected as a source follower. Recall that the input impedance at the gate of an FET is dominated by a capacitance, like the lumped-element electrodes of an MZM. Consequently the gate current is negligible, but the gate voltage can control a

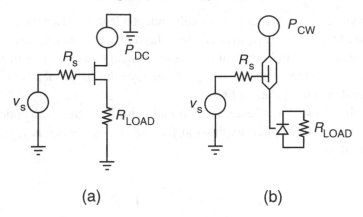

Figure 3.5 (a) Schematic of an electronic amplifier based on a field effect transistor (FET). (b) Block diagram of an externally modulated optical link.

large source-to-drain current. Consequently the modulated power in the source load can be larger than the modulation power drawn by the gate, resulting in modulated power gain between the gate and source. We can think of an external modulation link in an analogous fashion, where the MZM replaces the FET and the CW optical power replaces the FET drain-to-source voltage, as shown in Fig. 3.5(b).

The average optical power at which the direct and external modulation link gains are equal is not a fundamental constant since the power at which this occurs depends on the parameters we assumed for the calculation. Thus this power can be used as a measure of the relative state of the art between the two modulation techniques. The cross-over power, P_C, can be calculated by setting the two expressions for link gain, (3.7) and (3.10), equal, canceling the common terms and solving for P_C. The result is

$$P_C = \frac{2s_\ell V_\pi}{T_{FF}\pi R_S}. \tag{3.11}$$

The higher P_C the more efficient direct modulation is, and consequently the wider the range of optical powers is over which the gain of a direct modulation can exceed that of external modulation. Conversely the lower the cross-over power, the more efficient external modulation is.

The elucidation of the optical power dependence for link gain can also resolve one of the early debates in the development of analog links: given a diode laser, is it better to modulate it directly or to use this *same* laser as the CW source for an external modulator? The answer usually was that of course it is better to modulate directly, because the unavoidable loss of the modulator will always make external modulation have lower gain. However, as (3.7) and (3.10) show, this line of reasoning leads one away from a critical distinction. As these equations show, if the average optical power is fixed at a low value, direct modulation will always have higher gain than external, even if the modulator optical loss was zero. On the other hand, if the CW optical power is high, then external modulation will have a higher gain than direct, in spite of the additional modulator loss. By imposing the constraint that the same laser be used for both types of links, early workers had inadvertently, but fundamentally, cast the decision in favor of direct modulation, since early diode lasers typically had optical powers of less than 1 mW. More recently, significantly higher powers have become available, either from diode lasers or from diode laser pumped solid state lasers. Thus it is no longer necessary, or even desirable, to constrain the same type of laser to be used in both direct and external modulation. As we shall see in subsequent chapters, the flexibility in laser choice for external modulation has ramifications that go beyond the link gain.

With intrinsic gains for links being reported that rival the gains from traditional electronic amplifiers, it is natural to ask: could such links replace amplifiers? The answer is in principle yes, but in practice not yet. The primary reason is based on a factor we have not discussed yet: the gain efficiency, i.e. the dc power required to produce a given level of amplifier performance. The way we choose to provide a quantitative basis for this comparison[1] requires the use of the 1-dB compression power, which will be defined in Section 3.4. Therefore we defer discussion of this topic to that section.

3.3.2 Wavelength

Recall from Chapter 2 that the diode laser slope efficiency and photodiode responsivity expressions, equations (2.6) and (2.35) respectively, had a wavelength dependence. We now substitute these equations into (3.7); the result is

$$g_i = \left(\frac{\eta_l hc}{qn\lambda_0}\right)^2 \left(\frac{\eta_d q\lambda_0}{hc}\right)^2. \tag{3.12}$$

If we assume that the laser and photodiode are fabricated in materials with the same refractive index, then we can cancel the common terms in (3.12) to obtain

$$g_i = \eta_l^2 \eta_d^2 \le 1. \tag{3.13}$$

Thus in the case of direct modulation, the wavelength dependence of the individual components drops out to produce an expression for intrinsic gain that is independent of wavelength.

This form of the direct modulation link intrinsic gain also permits us to explore the limits on achievable link gain. If we continue to assume a regular PIN photodiode,

---

[1] Another possible basis for comparison would be the power added efficiency, PAE. We choose not to use PAE here because it is more applicable to high power amplification, as opposed to the low power amplification we are discussing here.

then its external differential quantum efficiency $\eta_d \leq 1$. To achieve intrinsic gain greater than one, this implies a laser external differential quantum efficiency $\eta_\ell > 1/\eta_d$. This is not possible with a conventional single section diode laser. However, there are diode lasers, such as the gain lever laser (Vahala *et al.*, 1989) and the cascade laser (Cox *et al.*, 1998), for which $\eta_\ell > 1$.

Conversely we could use a conventional single section laser and substitute a photodetection device with $\eta_d > 1$, such as an avalanche photodiode, for the PIN photodetector. As long as $\eta_d > 1/\eta_\ell$, the link will have an intrinsic gain greater than one. In fact, from a gain point of view, this and the previous case are indistinguishable. The distinction between increasing the laser slope efficiency and photodetector responsivity will become clear when we consider noise effects in links in Chapter 5.

To investigate the wavelength dependency of intrinsic gain for external modulation, we start by augmenting (3.10) with the explicit expression for photodiode responsivity, (2.35):

$$g_{\text{mzpd}} = \left(\frac{T_{\text{FF}} P_{\text{I}} \pi R_{\text{S}}}{2 V_\pi} \right)^2 \left(\frac{\eta_d q \lambda_0}{hc} \right)^2. \tag{3.14}$$

Unfortunately little insight is obtained from (3.14), since the modulator slope efficiency and photodetector responsivity are in different forms. We can remedy this situation by recognizing that P_{I} can be expressed as the energy per photon multiplied by the number of photons per second, p:

$$P_{\text{I}} = \left(\frac{hc}{\lambda_0} \right) p. \tag{3.15}$$

Substituting (3.15) into (3.14) and canceling the common terms yields

$$g_{\text{mzpd}} = \left(\frac{T_{\text{FF}} p \pi R_{\text{S}}}{2 V_\pi} \right)^2 (\eta_d q)^2. \tag{3.16}$$

From (3.16) it appears that the wavelength dependence has canceled out, just like the direct modulation case. However, recall from the discussion following (2.18) in Chapter 2 that $V_\pi \propto \lambda^2$ for two modulators that are identical in design, electro-optic material, etc., but optimized to operate at different wavelengths; consequently $g_{\text{mzpd}} \propto 1/\lambda^4$. The inverse fourth power scaling of gain with wavelength means that even a relatively small wavelength decrease can have a significant impact on link gain. For example, reducing the wavelength from 1.55 μm to 1.3 μm with all other link parameters held constant would increase the intrinsic gain by 3 dB. Changing the wavelength from 1.3 μm to 0.85 μm would increase the gain by another 7 dB.

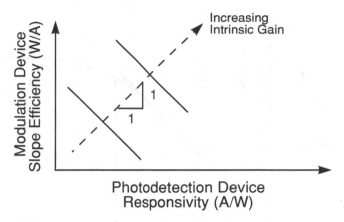

Figure 3.6 Contours of constant intrinsic link gain vs. the modulation slope efficiency and photodetection responsivity.

However, there are practical factors at present – such as photorefractive damage[2] in lithium niobate – that mitigate against using this fact to much advantage. Also, this conclusion does not generalize across all modulator types as the optical power dependence did.

3.3.3 Modulation slope efficiency and photodetector responsivity

If we examine the intrinsic gain expressions for both direct and external modulation, (3.7) and (3.9), we notice that the modulation slope efficiency and photodetection responsivity appear in the same functional form in both expressions. What this implies is that changes in either the slope efficiency or the responsivity are equally effective at changing the intrinsic gain. For example if we want to double the intrinsic gain, we can do so by increasing only the modulation slope efficiency by $\sqrt{2}$ or only the photodetection responsivity by $\sqrt{2}$ or by any combination of modulation slope efficiency and photodetection responsivity whose product equals $\sqrt{2}$.

To see this fact graphically, in Fig. 3.6 we plot contours of constant intrinsic link gain as a function of modulation slope efficiency and photodetection responsivity. The fact that the slope of these contours is one reflects the fact that either slope efficiency or responsivity is equally effective at increasing the intrinsic gain. This result may seem so obvious to some as to make one question why we are even raising the point. The answer is that we want to establish a basis for contrasting

---

[2] A small fraction of the optical power in the waveguides is absorbed, generating electron–hole pairs, which in turn establish an electric field across the waveguide. Because we are working in an electro-optic material, this electric field changes the index of refraction, thereby altering the bias point of the modulator. At 0.85 μm, 50 μW of optical power can cause measurable drift.

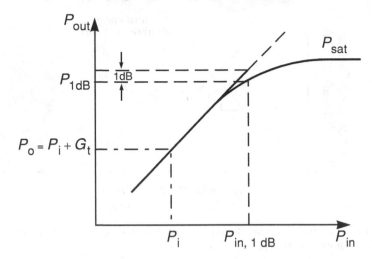

Figure 3.7 RF output power of a device with gain G_t as a function of RF input power showing gain compression and the 1-dB compression point.

the effectiveness of increasing slope efficiency vs. increasing responsivity on gain with the effectiveness of these same two parameters on improving the noise figure, which we will do in Chapter 5.

3.4 Large-signal intrinsic gain

In the small-signal regime, the gain is independent of input signal power; i.e. increase the input signal and we obtain the *same* gain. If we plot (Fig. 3.7) device output power vs. input power on log-log scales, independence of small-signal gain is represented by the straight line portion of this function. However, as we continue to increase the modulation power, we eventually notice that the gain begins to decrease. This reduction in gain from the small-signal value is called *compression*. Further increases in input signal power result in further decreases in gain until a maximum output power is reached. Beyond this, increases in input power produce *no* increase in output power. At this point the device is said to be *saturated*, which is indicated by the horizontal portion of the function in Fig. 3.7.

When we begin to enter compression, the accuracy of the small-signal model decreases. We investigate one consequence of this effect here: gain compression; we will investigate another consequence of this effect – distortion – in Chapter 6.

There are two general approaches to modeling gain compression. The first approach is general in that it can be applied to all links without any detailed knowledge of the individual modulation or photodetection device transfer functions. However, the price one pays for this generality is that it provides little, if any, insight into how to modify the device transfer function, should this be necessary to meet the application. The other approach addresses this shortcoming of the general approach by relating the slope efficiency decrease to the underlying device parameter(s). The disadvantage of this approach is that it needs to be redeveloped for each device, since the underlying gain compression mechanisms are typically device dependent.

For either approach we need a measure of gain compression. A commonly used value of gain compression is 1 dB (see for example Gonzalez, 1984); i.e.

$$G_{1\text{-dB}} = G_i - 1. \tag{3.17}$$

That is, the 1-dB compression gain is simply 1 dB less than the small-signal intrinsic gain.

An important figure of merit for RF components, including analog optical links, is the input signal level, P_{in}, which causes $G_{1\text{-dB}}$. Expressing the 1-dB compression point at the link output, $P_{1\text{-dB}}$, in terms of $G_{1\text{-dB}}$ yields

$$P_{1\text{-dB}} - P_{in} = G_{1\text{-dB}}, \tag{3.18}$$

where the RF powers are expressed in the same dB units, i.e. dBm, and the gain in dB.

The general approach to handling large-signals is based on the results of Section 3.2 where we saw that the intrinsic gain for any link can be expressed as the square of the modulation slope efficiency and photodetection responsivity. Consequently the general approach takes the modulation slope efficiency and photodetector responsivity as given and asks: at what modulation level does the product of the slope efficiency and responsivity decrease to the point where the intrinsic gain has decreased by 1 dB? Since the same derivation holds for any link, we choose to use direct modulation here and consequently start the development by taking the log of (3.7):

$$G_i = 20 \log s_\ell + 20 \log r_d. \tag{3.19}$$

Link gain compression is virtually always dominated by the modulation device. Expressing this observation analytically means that we subtract all the gain compression, i.e. a full 1 dB, from the first term on the right hand side of (3.19). Thus the 1-dB compression slope efficiency, in dB, can be written as

$$S_{\ell\text{-1dB}} = 20 \log s_\ell - 1. \tag{3.20}$$

Converting (3.20) out of dBs yields

$$s_{\ell\text{-1dB}} = 10^{(20 \log s_\ell - 1)/20}. \tag{3.21}$$

For a numerical example of (3.21), consider a diode laser with a slope efficiency of 0.2 W/A. Substituting this value into (3.21), the 1-dB compression gain occurs when the slope efficiency has decreased to 0.178 W/A.

As was mentioned above, the general approach is fine if, for example, one is trying to trying to meet a link specification by choosing from among a number of existing devices based on their published performance. However, one often finds in practice that none of the existing devices meets a particular need and so the question arises as to how the device slope efficiency needs to be modified to meet a given link specification. In such situations, the second approach introduced above can be used.

Like the general approach, the second approach can be applied to any link. Thus the second approach lacks generality only in the sense that it must be rederived for each device type. We choose to illustrate the second approach by using a Mach–Zehnder external modulation link.

Consider signals large enough to violate the small-signal approximation that we used to develop the slope and incremental modulation efficiencies in Chapter 2. Consequently to begin the large-signal derivation we need to go back to the basic equation for a Mach–Zehnder, (2.12), which is repeated below for convenience:

$$p_{M,O} = \frac{T_{FF} P_I}{2} \left(1 + \cos \left(\frac{\pi v_M}{V_\pi} \right) \right). \tag{3.22}$$

Next we need to assume a specific waveform for the large input signal. For at least three reasons a sine wave is a good choice: it is analytically tractable, experimentally convenient and universally generalizable since any other input can be decomposed into a sum of sine waves. We also need to assume a bias point; we will choose $V_\pi/2$ as this clearly permits the largest amplitude swing in the modulation voltage. Combining these assumptions results in a modulator voltage $v_M = (V_\pi/2 + v_m) \sin(\omega t)$, which we then substitute into (3.22):

$$p_{M,O} = \frac{T_{FF} P_I}{2} \left(1 + \cos \left(\frac{\pi}{2} + \frac{\pi v_m \sin(\omega t)}{V_\pi} \right) \right). \tag{3.23}$$

The standard way to proceed with the analysis of an equation that contains $\cos(\sin(x))$ terms is to do an expansion in terms of Bessel functions of the first kind of various orders, $J_n(x)$. This analytical approach for large-signal gain was reported for a Mach–Zehnder by Kolner and Dolfi (1987). The first two terms in the Bessel function expansion of (3.23), which represent the dc and gain at the fundamental, are

$$p_{M,O} = \left(\frac{T_{FF} P_I}{2} \right) (1 + 2J_1(\phi_m) \sin(\omega t) + \cdots), \tag{3.24}$$

where $\phi_m = \pi v_m / V_\pi$.

We are interested here in developing an expression for the large-signal transfer function for a Mach–Zehnder; consequently we only need the coefficient of the fundamental term in (3.24):

$$p_{m,o} = T_{FF} P_I J_1(\phi_m). \tag{3.25}$$

We proceed as we did with the small-signal development and express the large-signal modulation efficiency as the ratio of the square of (3.25) to the available source power, where we again assume that the modulator is impedance matched to the source, as we did in the small-signal case:

$$\frac{p_{m,o}^2}{p_{s,a}} = \frac{(T_{FF} P_l)^2 J_1^2(\phi_m)}{\dfrac{v_s^2}{4R_S}}. \tag{3.26}$$

Although (3.26) is usable in the sense that one could use it to calculate the 1-dB power, it is not very insightful because it obscures the relationship between the large and small-signal modulation efficiencies.

With the modulator impedance matched to the source, $v_m = v_0/2$. Therefore we can express v_s in terms of ϕ_m and V_π: $v_s = 2V_\pi \phi_m/\pi$. Substituting this expression for v_s into (3.26) and grouping the terms as we did for the small-signal modulation efficiency, (2.17), yields an alternate form of (3.26), the large-signal modulation efficiency for a Mach–Zehnder modulator:

$$\frac{p_{m,o}^2}{p_{s,a}} = \left(\frac{T_{FF} P_l \pi R_S}{2V_\pi}\right)^2 \frac{1}{R_S} \left(\frac{2J_1(\phi_m)}{\phi_m}\right)^2, \tag{3.27}$$

where as with the incremental modulation efficiency we have multiplied (3.26) by R_S/R_S.

The first term in (3.27) is just the small-signal slope efficiency for a Mach–Zehnder as given by (2.18) and in combination with the second term in (3.27) is the expression for the small-signal modulation efficiency as given by (2.17). Thus we can immediately rewrite (3.27) as

$$\frac{p_{m,o}^2}{p_{s,a}} = \frac{s_{mz}^2}{R_S} \left(\frac{2J_1(\phi_m)}{\phi_m}\right)^2. \tag{3.28}$$

The (3.28) form of the large-signal modulation efficiency makes clear the gain compression, provided we can show that for $0 < \phi_m < \infty$ the term in parentheses in (3.28) is ≤ 1.

The value of the term in parentheses in (3.28) at $\phi_m = 0$ is indeterminate since both the numerator and denominator are zero. A common technique to establish a value in such cases is to invoke l'Hôpital's rule (see for example Thomas, 1968), which states that if the ratio of the derivatives of the numerator and denominator exists as $\phi_m \to 0$, then

$$\lim_{x \to 0} \frac{f(x)}{g(x)} = \lim_{x \to 0} \frac{f'(x)}{g'(x)}. \tag{3.29}$$

Applying (3.29) to establish the value of the term in parentheses in (3.28) as $\phi_m \rightarrow 0$ yields

$$\lim_{\phi_m \rightarrow 0} \left(\frac{2\phi_m J_1(\phi_m)}{\phi_m^2} \right) = \lim_{\phi_m \rightarrow 0} \left(\frac{\dfrac{d(2\phi_m J_1(\phi_m))}{d\phi_m}}{\dfrac{d(\phi_m^2)}{d\phi_m}} \right) = \lim_{\phi_m \rightarrow 0} \left(\frac{2\phi_m J_0(\phi_m)}{(2\phi_m)} \right) = 1,$$

(3.30)

where we have multiplied the numerator and denominator of the term in parentheses in (3.28) by ϕ_m so that we can take the derivative of the Bessel function by using the identity (see for example Wylie, 1966): $d(\phi_m J_1(\phi_m))/d\phi_m = \phi_m J_0(\phi_m)$. In obtaining the final limit in (3.30) we also used the fact that $J_0(0) = 1$.

Consequently as $\phi_m \rightarrow 0$, (3.28) $\rightarrow$ (2.17), i.e. for small-signals the large-signal expression – (3.28) – reduces to the small-signal expression.

As an example, for $\phi_m < 0.06$, the Bessel function term in (3.28) contributes less than a 1% correction to the modulation efficiency calculation based on the small-signal assumption; i.e. the first term in (3.28) alone. For this value of ϕ_m, $v_m \cong 0.02\,V_\pi$. Stated in words, this means that a modulation voltage that is only 2% of the modulator half wave voltage is large enough to begin to introduce measurable gain compression. Consequently, one normally operates a Mach–Zehnder modulator at modulation voltages below this value – and when we get to considering the effects of distortion, we will see that the maximum modulation depth is even further reduced. These modulation depths for analog applications are in sharp contrast to the modulation depths used for digital, which typically approach 100%.

We can use (3.28) to calculate the input power that results in 1-dB gain compression by setting the Bessel function term in (3.28) equal to the numeric value that corresponds to 1-dB:

$$\frac{2J_1(\phi_m)}{\phi_m} = 10^{-\frac{1}{20}}.$$

(3.31)

The solution of (3.31) (Kolner and Dolfi, 1987) in terms of the peak voltage for 1-dB compression is

$$v_{1\text{-dB}} = \frac{V_\pi}{\pi}.$$

(3.32)

Converting peak voltage to RMS voltage and using the previous expressions relating v_m to v_s produces an expression for the input power for 1-dB compression for a Mach–Zehnder modulator:

$$p_{1\text{-dB,MZ}} = \frac{V_\pi^2}{2\pi^2 R_S}.$$

(3.33)

Table 3.2 *Comparison of RF amplifier and optical link gain efficiency*

Component	Gain (dB)	$P_{1\text{-dB}}$ (mW)	P_{DC} (mW)	E (%)
External modulation link	15	0.25	50 000	0.0005
($V_\pi = 0.5$ V; $P_I = 400$ mW)				
RF electronic amplifier	15	25	200	0.13

As an example, consider a Mach–Zehnder modulator with a $V_\pi = 5$ V matched to a 50 ohm source; then $p_{1\text{-dB,MZ}} = 25.3$ mW or 14 dBm. If we take 1% gain compression as the dividing line between the small- and large-signal modulation efficiencies, then we can use the small-signal expression up to input power levels of 91 μW or about −10 dBm.

The second approach to analyzing the effects of gain compression can be applied to any other modulation or photodetection device, by following the above procedure provided we have an analytic expression for the device transfer function. For example in the case of direct modulation of a diode laser, one would start with the diode laser gain compression factor (Coldren and Corzine, 1995).

Having established the 1-dB compression power, we now have the tools necessary to quantify the concept of gain efficiency that was introduced in Section 3.3.1. We will define the gain efficiency, E, of an amplifying component to be the ratio of its 1-dB compression power – which is an RF power – to the dc power, P_{DC}, drawn when producing that RF power,

$$E = \frac{p_{1\text{-dB}}}{P_{DC}}. \tag{3.34}$$

Table 3.2 lists the gain efficiencies as defined above for the externally modulated optical link presented in Section 3.3.1, along with the efficiency typical of present, commercially available RF electronic amplifiers. As can be seen from these numbers, an externally modulated optical link with high gain has substantially lower efficiency than does its electronic counterpart. The reason for the low efficiency can be seen by re-examining (3.10). Present links have achieved positive intrinsic gains primarily by increasing the average optical power. However, as (3.10) makes clear, it is really the ratio of average optical power to modulator sensitivity that determines the gain. Either factor is equally effective at *increasing* the gain, but increasing the modulator sensitivity is more *efficient* than increasing the average optical power in achieving a given value of gain.

Appendix 3.1 External modulation links and the Manley–Rowe equations

Conventional amplifiers basically achieve gain by arranging for a small-signal to vary the resistance connected across a dc (i.e. zero frequency) power source. It is also possible to achieve gain through the use of a non-linear reactive component and a RF power source. The most common type of parametric amplifier applies a large amplitude RF signal, the "pump", together with the signal it is desired to amplify, to a varactor diode. As its name implies, a varactor is a two terminal electrical component whose capacitance can be modulated by the reverse bias applied to it. The result of modulating the varactor capacitance is that a portion of the pump power is converted to produce an amplified version of the signal at the difference frequency between the pump and input signal frequencies. Parametric amplification has also been demonstrated in magnetic amplifiers by using non-linear magnetic materials.

In attempting to understand the properties of such amplifiers, an elegant general theory of amplification was developed by Manley and Rowe (1956, 1958) and generalized by Penfield (1960). One of the important conclusions of their work was that the maximum power gain from a parametric amplifier is set by the ratio of the pump to signal frequencies.

When the first externally modulated links with positive intrinsic gain were reported, it appeared to some researchers that gain from these links was not too surprising since they viewed them as merely a new form of parametric amplifier. Indeed the superficial similarity between these links and a parametric amplifier is striking: both have a modulator and both use a carrier for the power source. However, if we apply the Manley–Rowe formula for maximum gain to an externally modulated link we obtain 43 dB, assuming an optical carrier at 1.3 μm (231 THz) and a maximum signal frequency of 10 GHz!

Could this be true? The answer is yes and no: yes there is modulation gain, but no this gain is not accessible from a link perspective. The difficulty lies in the fact that in the case of an externally modulated link, this parametric modulation gain is followed by an equal parametric demodulation loss when the photodetector demodulates the optical carrier to recover the RF modulation. A similar situation occurs with magnetic amplifiers when one attempts to use a magnetic demodulator (Manley, 1951).

Consequently to realize the parametric modulation gain in a link, one would need an *optical* non-linearity to apply the Manley–Rowe model, which we clearly do not have in the case of an externally modulated link.

References

Coldren, L. A. and Corzine, S. W. 1995. *Diode Lasers and Photonic Integrated Circuits*, New York: John Wiley & Sons, 195–6.
Cox, C. H., III, Roussell, H. V., Ram, R. J. and Helkey, R. J. 1998. Broadband, directly modulated analog fiber link with positive intrinsic gain and reduced noise figure, *Proc. Int. Topical Meeting on Microwave Photonics 1998*, IEEE, 157–60.

Gonzalez, G. 1984. *Microwave Transistor Amplifiers Analysis and Design*, Englewood Cliffs, NJ: Prentice-Hall, Inc., 175–6.

Kolner, B. H. and Dolfi, D. W. 1987. Intermodulation distortion and compression in an integrated electrooptic modulator, *Appl. Opt.*, **26**, 3676–80.

Linvill, J. G. and Gibbons, J. F. 1961. *Transistor and Active Circuits*, New York: McGraw-Hill Book Company, Inc., 233–6.

Manley, J. M. 1951. Some general properties of magnetic amplifiers, *Proc. IRE (now IEEE)*, **39**, 242–51.

Manley, J. M. and Rowe, H. E. 1956. Some general properties of nonlinear elements – Part I. General energy relations, *Proc. IRE (now IEEE)*, **44**, 904–13.

Penfield, P. 1960. *Frequency Power Formulas*, New York: John Wiley & Sons.

Pettai, R. 1984. *Noise in Receiving Systems*, New York: John Wiley & Sons, Chapter 7.

Rowe, H. E. 1958. Some general properties of nonlinear elements – II. Small-signal theory, *Proc. IRE (now IEEE)*, **46**, 850–60.

Thomas, G. B. 1968. *Calculus and Analytic Geometry*, 4th edition, Reading, MA: Addison-Wesley Publishing Co., 651.

Vahala, K., Newkirk, M. and Chen, T. 1989. The optical gain lever: A novel gain mechanism in the direct modulation of quantum well semiconductor lasers, *Appl. Phys. Lett.*, **54**, 2506–8.

Wylie, C. R. Jr. 1966. *Advanced Engineering Mathematics*, 3rd edition, New York: McGraw-Hill Book Co., 365.

4

Frequency response of links

4.1 Introduction

The device slope efficiencies that we developed in Chapter 2, and that were cascaded to form links in Chapter 3, explicitly ignored any frequency dependence. In this chapter we remove that restriction. As we shall see, virtually all modulation and photodetection devices have an inherently broad bandwidth. Digital links require broad bandwidth, which is one of the reasons for the numerous applications of fiber optic links to digital systems. A few analog link applications also require the full device bandwidth. However, it is far more common for analog links to need only a portion of the devices' inherent bandwidth. Consequently most analog link designs include some form of RF pre- or post-filtering to reduce the bandwidth.

For completeness we address bandpass and broad bandwidth impedance matching for three electro-optic devices: PIN photodiode, diode laser and Mach–Zehnder modulator. We then combine the bandpass impedance matched cases to form both direct and external modulation links. However, the same analytical approach is used for both impedance matching methods and both modulation techniques. Therefore those readers desiring a less exhaustive treatment can obtain a complete introduction to the subject by studying only one of the impedance matching methods and one of the modulation techniques.

One may be tempted to ask: why bother with bandwidth reduction, since this adds components and complicates the design? There are at least two key reasons for implementing bandwidth reduction. One, which is explored in this chapter, derives from the fact that we can trade excess bandwidth for increased intrinsic gain, i.e. lower RF loss in most present links. In turn, lower loss permits lower link noise figures, as we will see in Chapter 5. The other key reason for implementing bandwidth reduction is that it expands the trade space for distortion reduction techniques, which as it turns out also involves noise figure. We will explore these aspects of bandwidth reduction in Chapters 6 and 7, respectively.

The requisite bandwidth reduction can, of course, be achieved with either active or passive electronic components. Most active devices introduce gain, in addition to any bandwidth reduction they may perform. Thus such devices can obscure the changes in the intrinsic gain that are due solely to the amount of bandwidth reduction. Therefore we choose to focus on passive components here since this makes clearer the tradeoffs between link parameters and the degree of bandwidth reduction. Active devices can – and usually do – introduce noise and distortion as well, which would obscure the intrinsic link noise and distortion, respectively. Thus the approach we are taking in this chapter will also be used in the next two chapters.

Recall from Chapter 2 that the definition of link gain requires that there be impedance matches between the source/load and the modulation/photodetection devices. Without any frequency dependence, this requirement was met with a resistor. As one might expect, expanding the modulation/detection devices to include frequency dependence requires that we expand the impedance match to include its frequency dependence as well. It turns out that the frequency dependent matching circuit is intimately related to, and in some cases is synonymous with, the bandwidth reduction circuit.

Rarely does the link designer have the luxury of having a custom electro-optic device designed for a particular application. More commonly the link designer needs to select an existing device and design the RF filter to work with the selected device. The organization and depth of the material in this chapter reflects this latter approach. In Section 4.2, the factors that affect the frequency response of the more common electro-optic devices are briefly reviewed. Section 4.2 is not intended to be sufficiently detailed that the reader can design an electro-optic device, but rather is intended to give a general rationale as to why the device frequency response is what it is.

On the other hand, the discussion in Section 4.2 is intended to be sufficiently detailed so as to permit the reader to design the desired filter/matching circuit. This is the focus of Section 4.3, where we investigate the impacts of broad- and narrow-band impedance matching on intrinsic gain. To do so requires that we expand the device models presented in Chapter 2 to include the first-order frequency response effects of the various modulation and photodetection devices. This material is also presented in Section 4.3.

As one might expect, there is a relationship between the quality of the impedance match and the bandwidth over which one can maintain the quality of the match. This relationship is formalized by the Bode–Fano limit and is discussed in Section 4.4.

One point on conventions is important to make before we begin these discussions. It is common in electrical engineering to quote bandwidth in terms of the frequency at which the electrical output is 3 dB down from some reference value, such as the mid-band or low frequency value. When the bandwidth of the electro-optic devices is viewed from a *link* perspective – i.e. on an electrical in, electrical out basis – there is no ambiguity in how to apply the traditional electrical engineering definition of bandwidth. However, electro-optic device designers often do not construct an entire link when characterizing their latest device but rather measure the electro-optic *device* bandwidth. Consequently they often quote the "3-dB bandwidth" at the frequency where the *optical* response is down by 3 dB. As we saw in Chapter 3, the electrical response of a link is the square of the optical response; consequently the *electrical* response would be down 6 dB at the optical 3-dB bandwidth.

To obtain the exact relationship between the frequencies for these two bandwidth definitions requires an evaluation of the frequency response involved. However, the high frequency portion of the transfer function of many electro-optic devices can be approximated by the low pass frequency response of a distributed system:[1]

$$\left(\frac{1}{\sqrt{\tau s}} \right),$$
(4.1)

where τ is a characteristic time constant of the device, and $s = \alpha + j\omega$ denotes the complex frequency. Using this approximation, it is clear that the relationship between the electrical and optical 3-dB bandwidths is

$$s_{3\text{-dB,elect}} = \frac{s_{3\text{-dB,opt}}}{\sqrt{2}} \approx 0.7 s_{3\text{-dB,opt}}.$$
(4.2)

Thus when discussing bandwidth it is important to ascertain the domain, electrical or optical, to which the author is referring. In the following, all bandwidths are quoted in terms of their electrical 3-dB bandwidth.

Finally, we return to the small-signal assumption for all the material presented in this chapter.

4.2 Frequency response of modulation and photodetection devices

4.2.1 Diode lasers

In Chapter 2 we discussed three types of diode laser cavities: Fabry–Perot, distributed feedback, and vertical cavity, surface emitting. However, we found that all three laser structures shared a common dc/low frequency model. It turns out that all three laser structures also share a common dynamic model, although like their low frequency counterpart, the specific values for the various parameters that go into the dynamic model for each cavity type depend on the laser cavity design.

---

[1] Throughout this chapter we use the frequency domain description of the filtering/matching circuits. For a review of this material, see for example Van Valkenburg, Chapter 7 (1964).

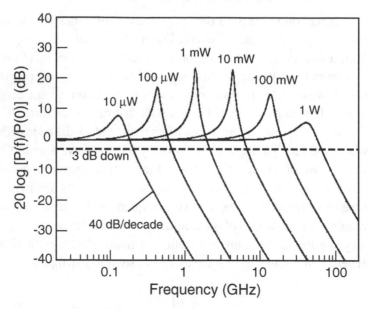

Figure 4.1 Plot of the calculated frequency response for an idealized high frequency diode laser with the laser output power as a parameter. (Coldren and Corzine, 1995 Fig. 2.12. © 1995 John Wiley & Sons, Inc. This material is used by permission of John Wiley & Sons, Inc.)

A typical diode laser frequency response for small-signal amplitude modulation is plotted in Fig. 4.1. As shown, the modulation response is relatively uniform over a wide frequency range. The response peaks at the upper end of the range and then decreases rapidly with frequency above the peak. Further, the frequency of the peak response increases as the laser bias is increased above threshold.

Like the laser P vs. I curve, the laser frequency response also can be modeled quite accurately using the rate equations. However, we choose not to pursue that formalism in the body of this chapter because the results are in terms of myriad fundamental laser parameters, such as the bimolecular recombination coefficient, that are not directly relevant to the link designer. Instead we develop the dependences needed for link design, based on plausibility arguments in this section. For a more extended, and more rigorous, derivation of laser frequency response in terms of the rate equations, the reader is referred to Appendix 4.1 where we extend the dc rate equations developed in Appendix 2.1.

The peak in the frequency response is suggestive of a resonance and in fact is referred to as the *relaxation resonance*. A resonance in turn implies that there must be at least two coupled sources of energy storage. In a laser one reservoir of stored energy is the collection of injected carriers, from which the stimulated emissions are drawn. The other primary reservoir of stored energy in a laser is the collection of photons in the laser cavity. These two pools of stored energy are clearly coupled since the stimulated photons in the laser cavity come from the pool of injected carriers. If we apply a step increase in the number of injected carriers – via a step increase in the laser bias current – then we increase the probability of a stimulated emission. As the number of stimulated emissions increases, the number of injected carriers begins to be depleted, at least until the injected carrier population can be restored.

The injected carriers and the cavity photons have characteristic lifetimes that represent their various decay mechanisms. The two primary mechanisms for injected carrier decay are conversion to stimulated emissions and non-radiative recombination, i.e. recombination that does not involve the generation of a photon. The primary mechanisms for photon decay are emission out of the cavity and absorption within the cavity. A rate equation development shows that the frequency of the relaxation resonance, f_{relax}, is inversely proportional to the geometric mean of the lifetimes of the injected carriers, τ_n, and the cavity photons, τ_p:

$$f_{relax} \propto \frac{1}{\sqrt{\tau_n \tau_p}}. \tag{4.3}$$

For diode lasers, the injected carrier lifetime is typically a few nanoseconds, which is much shorter than it is in other types of lasers. The lifetime of cavity photons is on the order of a few picoseconds in diode lasers, which is also shorter than in other types of lasers – primarily due to the higher optical gain of diode lasers. The higher optical gain permits diode laser cavities to be much shorter than those of other types of lasers: diode laser cavities are typically tens or hundreds of micrometers long, versus the centimeter to meter length cavities of other types of lasers. Further, the higher optical gain also permits a higher percentage of the cavity photons to be emitted on each pass; typically 68% of diode laser cavity photons are emitted per pass, whereas other types of lasers often emit only 1% of the cavity photons per pass.

As Fig. 4.1 also makes evident, the relaxation resonance frequency of a diode laser, and hence its modulation bandwidth, increases as the bias current increases. A plausibility argument for this effect begins with the fact that as the bias current increases, the optical output from the laser also increases. The increased optical output arises because the increased bias current provides more injected carriers and hence more opportunities for stimulated emissions. The increase in the number of cavity photons triggers more stimulated emissions, thereby reducing the average lifetime of an injected carrier. In turn the reduced lifetime of the injected carriers increases the relaxation resonance frequency. A rate equation development of this effect reveals two additional important facts about the relaxation resonance frequency dependence on laser bias. One fact is that it is not the absolute laser bias, but rather the laser bias above lasing threshold that is important. The other fact is that the increase in relaxation resonance is proportional to the square root of the bias above threshold. We can now refine (4.3) to include the laser bias dependence of relaxation resonance frequency:

$$f_{\text{relax}} = \frac{\sqrt{\dfrac{I_L}{I_T} - 1}}{\sqrt{2\pi \tau_n \tau_p}}. \tag{4.4}$$

If we substitute $\tau_n = 1$ ns, $\tau_p = 1$ ps and $I_L = 3I_T$ into (4.4) the result is $f_{\text{relax}} \cong 17.8$ GHz. This is a typical relaxation resonance for a diode laser; the range is from about 1 GHz to about 40 GHz (Weisser *et al.*, 1996). All other types of lasers have relaxation frequencies less than 1 MHz; consequently diode lasers are the only lasers with sufficient bandwidth for practical direct modulation applications.

To achieve maximum bandwidth, it is necessary to operate the diode laser at a bias far above laser threshold. In the 40 GHz example of relaxation resonance quoted above, the laser was biased at 20 times threshold. Although it would appear from (4.4) that in principle one could obtain an arbitrarily high relaxation resonance with a sufficiently high laser bias current, in fact the upper frequency is limited by a combination of theoretical and practical constraints. The primary theoretical constraints arise from non-linear gain and various carrier effects, all of which are ultimately related to the damping factor (Weisser *et al.*, 1996). The primary practical limits are imposed by circuit parasitics between the laser and the test fixture as well as by thermal heating of the laser due to the large laser bias currents required for high frequency modulation. Consequently, when choosing the laser bias current for a particular link design, the final bias decision is usually a tradeoff between bandwidth and other laser parameters, such as the decrease in slope efficiency with increasing bias above threshold.

Throughout the above discussion, it was assumed that the laser design, i.e. its structure, dimensions, material, etc., were all held constant. Changing any, or all, of these factors in general changes the relaxation resonance frequency.

It is important to note that the definition of modulation bandwidth commonly used by developers of diode lasers is the frequency at which the modulation response is 3 dB below the low frequency response – on the high side of the relaxation resonance peak. Thus this definition does not say anything about the flatness of the modulation response across the modulation bandwidth. As can be seen from Fig. 4.1, the resonance peak can be, and often is, more than 3 dB above the low frequency response value. Consequently, while the relaxation resonance sets the *maximum* bandwidth, other factors such as noise, distortion and modulation response uniformity may limit the *usable* bandwidth to a frequency considerably less than the relaxation resonance frequency.

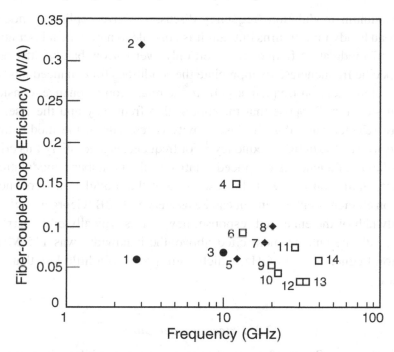

Figure 4.2 Plot of the fiber-coupled laser slope efficiency vs. maximum modulation bandwidth for the same lasers that were plotted in Fig. 2.3. The slope efficiencies have all been normalized to 1.3 μm. ● VSCEL, □ FP, ◆ DFB.

In addition to the dependence of modulation bandwidth on carrier lifetime, photon lifetime and bias current, there is also an important relationship between a laser's bandwidth and its slope efficiency. However, unlike the previously discussed dependences, the bandwidth/slope efficiency relationship is empirical; it represents the state of the art, not fundamental physics. To get a feel for this relationship, we again refer to the reports of demonstrated laser performance that were used to make the plot of fiber-coupled slope efficiency vs. wavelength, Fig. 2.3. For the plot of slope efficiency vs. maximum modulation frequency, we need to normalize out the effects of wavelength, since not all the results are for lasers operating at the same wavelength. Figure 4.2 is a plot in which all the slope efficiency data have been normalized to a wavelength of 1.3 μm.

As we can see from this plot, in general the diode laser slope efficiency decreases significantly as the maximum modulation frequency increases. The decrease is due in large part to the fact that it becomes increasingly difficult to modulate the laser current efficiently. The impact of the slope efficiency decrease is even more dramatic when viewed from the perspective of link gain, since the gain is proportional to the square of the slope efficiency.

The maximum modulation frequency discussed above applies to modulation over a broad bandwidth. It turns out that it is possible to modulate a laser above its "maximum" modulation frequency, albeit only over narrow bandwidths centered around specific frequencies. To appreciate the conditions for enhanced modulation efficiency at modulation frequencies above the relaxation resonance, consider the following situation. Suppose that the modulation frequency and the laser cavity length are selected such that the laser cavity is resonant at the modulation frequency *as well as* the optical frequency.[2] For frequencies that satisfy this criterion, the modulation efficiency is enhanced relative to the broadband modulation efficiency. Depending on the Q of the resonance at the modulation frequency, this modulation efficiency enhancement can be as great as 40 dB (Georges *et al.*, 1994). The bandwidth of the enhanced response, however, is typically quite narrow. For example, with the enhancement cited above, the bandwidth was 115 MHz at a modulation frequency of 45 GHz, which corresponds to a little less than a 0.3% bandwidth.

4.2.2 External modulators

As was discussed in Chapter 2, by far the most common modulator in use at present is a Mach–Zehnder interferometer fabricated in lithium niobate. Consequently this type of modulator is the most extensively modeled, and it is the focus of the discussion here. The directional coupler modulator, which is also predominantly fabricated in lithium niobate at present, has received far less attention in the literature. The design issues of its dynamic model are fundamentally the same and hence it is covered by inference under the Mach–Zehnder discussion. The third type of modulator that was introduced in Chapter 2 was the electro-absorption, EA, modulator. The frequency response of an EA modulator is basically determined by a set of tradeoffs that are similar to those of a photodiode; hence the EA modulator dynamic model is discussed in the next section.

The maximum modulation frequency of the electro-optic effect in lithium niobate is in the hundreds of GHz. Thus in practice the upper limit on the modulator frequency response is set not by the material but by the modulator design. This is in contrast with the limitations on maximum diode laser modulation frequency, where the limit is dependent on a combination of material and device parameters.

Also unlike the diode laser, a modulator on lithium niobate has no junctions; consequently the impedance, and hence the frequency response, is independent of modulator bias.

---

[2] This is perhaps more commonly known as the mode locking frequency.

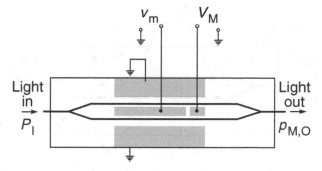

Figure 4.3 Mach–Zehnder modulator designed for low frequency modulation via lumped-element electrodes.

The key parameter that characterizes a Mach–Zehnder modulator's frequency response is the ratio of the optical transit time past the electrodes relative to the modulation period of the maximum modulation frequency. At relatively low modulation frequencies, the optical transit time is much shorter than the period of the maximum modulation frequency. Therefore the modulation signal is essentially constant during the optical transit time. In this case the electrodes can be treated as *lumped* elements as shown in Fig. 4.3. This is basically the modulator configuration that was shown in Fig. 2.7(a) but with separate electrodes for bias and modulation, which is a common alternative to using a single electrode for both bias and modulation. In this case the upper 3 dB frequency is simply determined by the RC roll-off due to the electrode capacitance and matching resistance, R_{MATCH}. For typical electrode spacing on lithium niobate, the electrode capacitance is about 0.5 pF per mm of electrode length. Consequently for a modulator with 10 mm electrodes and $R_{MATCH} = 50 \ \Omega$, the upper 3-dB frequency is 637 MHz because $\tau = 1.57$ ps.

This points out an important design tradeoff using lumped-element electrode modulators: to improve the link gain, one wants the lowest V_π possible, which in turn means longer electrodes. However, this increases the electrode capacitance, which in turn reduces the bandwidth. One way to resolve these opposing design constraints – when only a relatively narrow bandwidth is required – is to resonate out the electrode capacitance. When this is not an option, then usually one needs to go with traveling wave electrodes.

At relatively high frequencies, where the period of the maximum modulation frequency is not significantly greater than or is less than the optical transit time, lumped-element electrodes become less effective because the modulation voltage is not constant during an optical transit time. Continuing the modulator example from above, it takes the light in a waveguide about 73 ps to travel past a 10 mm electrode. This is about 10% of the period of a 1.4 GHz sine wave.

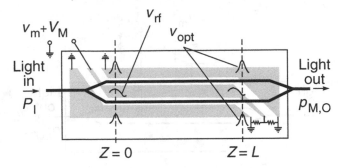

Figure 4.4 Mach–Zehnder modulator designed for high frequency modulation via traveling wave electrodes.

A common way to accommodate high frequencies in a distributed structure is to replace the lumped electrodes with traveling wave electrodes as shown in Fig. 4.4. In principle, the electrodes would be designed such that the RF propagation velocity along the traveling wave electrodes, v_{rf}, matches the optical propagation velocity in the waveguides, v_{opt}. In practice, it requires special measures to achieve this velocity match condition in lithium niobate while simultaneously achieving a characteristic impedance of 50 Ω.

The basis for the difficulty is the large difference between the dielectric constants of lithium niobate at RF and optical frequencies. The waveforms superimposed on the Mach–Zehnder structure shown in Fig. 4.4 illustrate this point. Suppose that the RF modulation starts out at the input end of the modulator such that its peak is aligned with the peak of the CW light feeding the modulator. Due to the difference in the optical and RF propagation velocities, the RF and optical peaks are not aligned at the output end of the modulator. In the extreme case illustrated in Fig. 4.4, the combined effects of the RF frequency and the difference in propagation velocities can result in the modulation imposed over the first half of the electrode length being completely canceled out by modulation of the opposite polarity over the second half of the electrode!

A commonly used figure of merit for the effects of velocity mismatch is the (electrode length)(bandwidth) product. Letting L represent the electrode length and B the electrical 3-dB bandwidth, the LB product can be expressed as (Betts, 1989)

$$LB = \left(\frac{1}{v_{rf}} - \frac{1}{v_{opt}} \right)^{-1} = \frac{c}{n_{rf} - n_{opt}}. \tag{4.5}$$

Substituting into (4.5) the common values for lithium niobate: $n_{rf} = 4.2$; $n_{opt} = 2.15$, the modulator response would be zero for $LB = 146$ GHz mm. A more practical value is where the electrical response is down by 3 dB from its value at dc; this turns out to be $LB = 64$ GHz mm. This means that a modulator with a 20 mm traveling wave "standard" electrode, i.e. an electrode without design features aimed specifically at velocity matching, would have a bandwidth of 3.2 GHz.

One might first think of solving this problem by simply making the electrodes shorter. However, without a commensurate increase in the electro-optic coefficient, the increased modulation bandwidth would come at the expense of decreased modulator sensitivity, manifested as a larger V_π. Thus the goal of high frequency modulator design is to increase the maximum modulation frequency without decreasing the modulator sensitivity.

A little further thinking might lead one to suggest simply removing some of the high dielectric lithium niobate, which should increase the RF propagation velocity. The initial difficulty with this approach is that, unlike semiconductors, lithium niobate is almost totally inert chemically, which makes virtually all etching slow at best. Early approaches to velocity matching searched for methods that would accommodate the dielectric differences while avoiding the fabrication difficulties. An example of such an approach (Alferness et al., 1984) is to accept the velocity mismatch but to avoid its deleterious effects by periodically reversing the electrode polarity whenever the difference in phase between the electrical and optical fields becomes too great. Since the phase reversal length is set at fabrication, the effectiveness of the phase reversal approach decreases the further one moves away from the design center frequency. Consequently a phase reversal modulator has a bandpass frequency response. The highest frequency results reported for a phase reversal modulator were a maximum modulation passband of 17.5–32.5 GHz at a V_π of 12 V (Wang et al., 1996). Thus this approach only partially achieved the desired goal.

Several groups (Noguchi et al., 1994; Gopalakrishnan et al., 1992a; Dolfi and Ranganath, 1992) have demonstrated true velocity matched modulators without significantly increased V_π. The layout of Noguchi et al.'s modulator is shown in Fig. 4.5(a). The key to their success was their ability to do pattern etching, albeit limited (only 3.3 μm deep), of the lithium niobate. When this etching process was combined with a judicious combination of design parameters they were able to achieve a V_π of 5 V and an upper 3-dB frequency of about 52 GHz – where we have used (4.2) to convert the reported optical bandwidth of 75 GHz to its corresponding electrical bandwidth. The frequency response of this result is shown in Fig. 4.5(b).

Once a velocity match is achieved, it is theoretically possible to have an arbitrarily low V_π by simply making the electrodes longer. However, there are presently three practical limitations on the maximum electrode length. The first is the RF loss of the electrodes. Virtually all electrodes are fabricated by evaporating gold a micrometer or so thick, which is only slightly thicker than the skin depth of gold at 10 GHz (about 0.7 μm). This loss can be reduced by making the electrodes thicker, as has

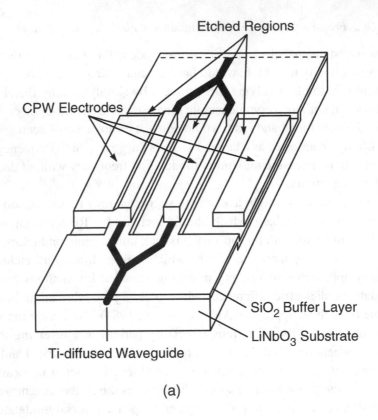

Etched Regions

CPW Electrodes

SiO₂ Buffer Layer

LiNbO₃ Substrate

Ti-diffused Waveguide

(a)

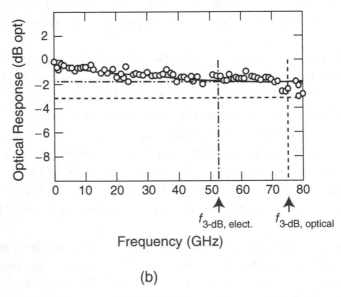

(b)

Figure 4.5 (a) High frequency Mach–Zehnder modulator with traveling wave electrodes. (b) Frequency response of a traveling wave Mach–Zehnder modulator. (Noguchi, K., Miyazawa, H. and Mitomi, O., 1994. 75 GH$_z$ broadband Ti:LiNbO$_3$ optical modulator with ridge structure, *Electron. Lett.*, **30**, 949–51. Reprinted with permission.)

been done by Gopalakrishnan *et al.* (1992b). They have developed a process for plating gold electrodes more than 10 μm thick, which although this does help reduce the loss it is still not thick enough to completely eliminate the losses due to skin effect and surface roughness.

The second present practical limitation on maximum electrode length is the size of the electro-optic crystal. Lithium niobate wafers are commonly available in 3″ diameter, with 4″ wafers beginning to appear. Almost 1″ is typically relegated to end face polishing, photolithographic edge effects and interferometer "Y" branches, leaving about 2″ as the maximum electrode length on a 3″ wafer (1″ ≈ 25 mm). With polymers, size is not an issue; consequently the limitation on maximum electrode length in polymers is not set by available material size.

The third practical limitation on maximum electrode length is the loss of the optical waveguides. If the optical loss is low, for instance 0.1 dB/cm as it is in lithium niobate, then 10 cm, or about 4″, of waveguides would only contribute 1 dB of optical loss to the modulator. Therefore waveguide loss is not typically a limiting factor on maximum electrode length in lithium niobate. However, when the optical loss is higher, for example 1 dB/cm, as it is in gallium arsenide and the polymers, then the same 10 cm of waveguides would contribute 10 dB to the overall optical loss of the modulator. Consequently the optical waveguide loss presently limits the maximum electrode length in these materials.

To get a picture of the slope efficiency–bandwidth tradeoff, as well as the present state of the art for Mach–Zehnder external modulators, we plot the fiber-coupled slope efficiency vs. upper 3-dB frequency using the same Mach–Zehnder modulators that were included in Fig. 2.10. These data are shown in Fig. 4.6. As with the analogous plot for diode lasers, all slope efficiencies have been normalized to operation at 1.3 μm. However, to make a comparison plot like this for external modulators also requires that we assume that all modulators were operated at the same average optical power. (Recall this fact from the definition for slope efficiency, (2.18).) For the slope efficiencies plotted in Fig. 4.6, we have assumed $P_I = 400$ mW, which is the highest power used at present in lithium niobate with only minor optical damage effects.

The data shown in Fig. 4.6 reveal three important points. One is that higher slope efficiencies have been obtained with external modulators than with directly modulated diode lasers. At low frequencies, it has been possible to achieve slope efficiencies large enough to yield positive intrinsic gains, i.e. links with power gains. Another important point is that there is roughly a f^{-1} dependence between the slope efficiency and the upper 3-dB frequency. This effect occurs despite the fact that any given modulator was optimized for the maximum slope efficiency at a given bandwidth. Thus this correlation is to be contrasted with the case where a particular modulator is operated at a variety of bandwidths. The third salient point

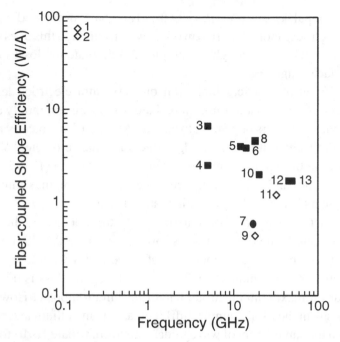

Figure 4.6 Plot of the modulator slope efficiency vs. maximum modulation band-width (based on the 3-dB electrical bandwidth) for the same modulators that were plotted in Fig. 2.10. The slope efficiencies have all been normalized to 1.3 μm and 400 mW of average optical input power. ● 830 nm, ◊ 1300 nm, ■ 1550 nm, □ 1550 nm (estimated from 3-dB optical BW).

from Fig. 4.6 is that higher modulation frequencies and bandwidths have been demonstrated with external modulators than with diode lasers.

As was the case with diode lasers, a few attempts have been made at trading a reduced bandwidth for a higher slope efficiency. Making the modulator electrode resonant at the modulation frequency is one approach that has been tried (Izutsu, 1996). Izutsu's initial results demonstrated a $V_\pi = 1.6$ V with a center frequency of 30.2 GHz.

As the frequency increases, it becomes increasingly difficult to couple the modulation signal to the electrodes. An elegant solution to this problem is to use radiative coupling of the modulating signal to the electrodes (Bridges *et al.*, 1991). As shown in Fig. 4.7, the modulator electrode consists of a series of dipole antennas. Coupling is accomplished when these dipoles are "illuminated" by the modulation signal from a waveguide. By segmenting the electrode and angling the modulator crystal with respect to the end of the waveguide, velocity matching also can be achieved. Bridges *et al.* demonstrated bandpass modulation around 94 GHz using this technique.

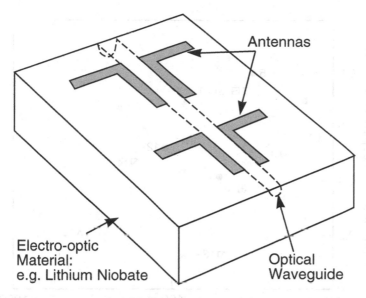

Figure 4.7 Antenna coupled Mach–Zehnder modulator. (Bridges *et al.*, 1991 © IEEE, reprinted with permission.)

There are two important notes to make in closing the discussion on the frequency response of Mach–Zehnder modulators. One is that, unlike the diode laser, there is no internal resonance mechanism with Mach–Zehnders. Therefore for these modulators, the usable bandwidth is the full 3-dB bandwidth of the modulator. The other note is that there is no optical power dependence to the Mach–Zehnder frequency response, as there is with the diode laser. As we recall from Chapter 3, one often chooses to operate an external modulation link at high average optical powers to improve the intrinsic gain of the link. Consequently there are reasons to operate both direct and external modulation links at high average optical powers, but one is driven to this common condition by fundamentally different reasons for each type of modulation.

4.2.3 Photodetectors

The only type of photodetector specifically discussed in Chapter 2 was the positive-intrinsic-negative, or PIN, photodiode. Consequently that is the focus of this section. However, we have also deferred to this section – from the previous section – the discussion of the EA modulator. The reason for combining the bandwidth discussion of a modulator and photodiode is the fact that like a PIN photodiode, EA modulators have a p-n junction that is reverse biased under normal operation, as opposed to a laser diode, for example, which is forward biased under normal operation. Thus

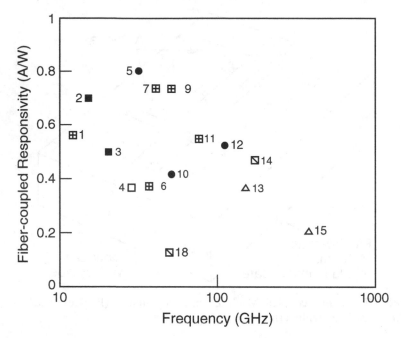

Figure 4.8 Plot of photodetector fiber-coupled responsivity vs. upper 3-dB frequency for the same photodetectors that were plotted in Fig. 2.15. • PIN (1500 nm), ◩ waveguide (830 nm), □ waveguide (1000 nm), ■ waveguide (1300 nm), ⊞ waveguide (1500 nm), △ MSM/Schottky (600 nm).

for purposes of developing its dynamic model, the EA modulator can be thought of as a PIN photodiode with controllable absorption that absorbs some, but not all, of the light incident on it.

The frequency response of a PIN photodiode is basically flat from dc up to an upper 3-dB frequency and then falls off, typically dominated by a single time constant. Thus like the Mach–Zehnder, a photodetector's usable bandwidth is its 3-dB bandwidth. However, like the diode laser, a photodetector's frequency response is the result of a combination of material and device effects.

Figure 4.8 is a plot of reported photodetector fiber-coupled responsivities vs. upper 3-dB frequency, using the same photodetectors whose responsivities vs. wavelength were plotted in Fig. 2.15. We can see two important points from Fig. 4.8. One is that photodetectors with frequency responses up to 500 GHz have been demonstrated, which is more than 10 times greater than the maximum modulation bandwidth of diode lasers and a factor of about 5 times greater than the maximum modulation bandwidth of external modulators. The other point is that the increases in upper 3-dB frequency come at the expense of lower responsivity. We now present a brief discussion on the basis for this tradeoff.

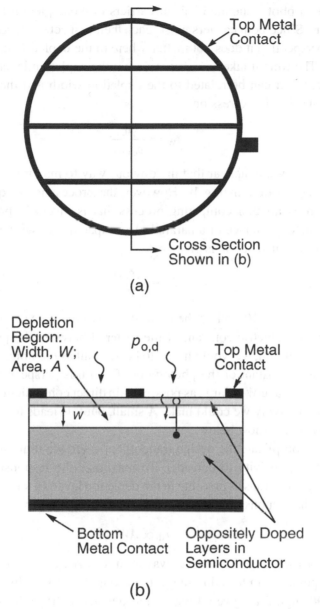

Figure 4.9 (a) Top and (b) cross section sketches of the basic layout of a surface illuminated PIN photodiode.

Consider the basic structure of a surface-illuminated PIN photodiode shown in Fig. 4.9. The bandwidth–responsivity tradeoff in this structure centers on the width, W, of the region that is depleted of carriers. The depletion width is basically the width of the intrinsic layer, but it does extend slightly into the doped layers when the photodiode is reverse biased. One of the bandwidth limiting effects is the time it takes a photogenerated pair of carriers in the depletion layer to reach the doped layers. Since both carriers must reach their respective doped layers, from an *efficiency* perspective it does not matter where in the depletion layer the photon is absorbed.[3] The time it takes a carrier to cross the depletion layer is known as the transit time, t_W; it can be related to the depletion width and the carrier velocity, $v_{carrier}$, via the simple expression:

$$t_W = \frac{W}{v_{carrier}}. \tag{4.6}$$

From (4.6) it would appear that the obvious way to maximize the bandwidth is to decrease t_W by minimizing W. However, the process is not quite that simple, because there is another competing process that is inversely proportional to W. Recall that the capacitance of a parallel plate capacitor, C, with electrodes of area A separated by a distance W is given by

$$C = \frac{\varepsilon_r \varepsilon_0 A}{W}, \tag{4.7}$$

where $\varepsilon_0 = 8.854 \times 10^{-12}$ F is the permittivity of free space and ε_r is the relative permittivity or dielectric constant of the material between the plates. Clearly for maximum bandwidth, we would like C to be as small as possible, so as to minimize τ, the RC time constant of the photodiode. From this perspective, (4.7) suggests that we want to make W as large as possible – in direct contradiction to the demands of (4.6). Alternatively we could make A small, but this leads to a conflict with the responsivity, as we show below.

To further complicate the design tradeoffs of a surface illuminated photodiode, there is the responsivity to consider. To maximize the responsivity, we want to absorb as many photons as possible in the depletion layer. In terms of device design parameters, the responsivity is proportional to the depletion volume, viz.:

$$r_d \propto AW. \tag{4.8}$$

Expression (4.8) suggests that the way to maximize r_d is to make the AW product as large as possible. In turn this suggests making A and W individually as large as possible, although for a given A one can also increase the responsivity by improving the fiber-to-photodetector coupling. However, making A large conflicts with minimizing τ (see (4.7)) and increasing W conflicts with minimizing the transit time t_W (see (4.6)), although it is consistent with minimizing C (see (4.7)). Given

---

[3] The velocities of the two types of carriers are typically different. Consequently the time it takes both carriers to reach the doped layers in general is a function of where the photon is absorbed. Therefore the *bandwidth* is a function of where the photon is absorbed.

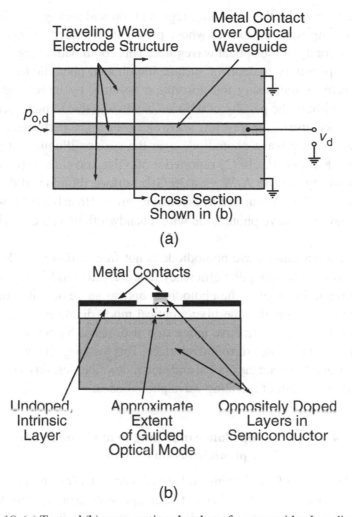

Figure 4.10 (a) Top and (b) cross section sketches of a waveguide photodiode with traveling wave electrodes.

all these conflicting design constraints, it is remarkable that designers of surface illuminated photodiodes have been able to produce the level of performance they have.

The traveling wave photodiode structure is an alternative to the surface illuminated photodiode that eases several of its conflicting design constraints. A sketch of a traveling wave photodiode is shown in Fig. 4.10. The key feature to the design flexibility of this type of photodiode is that the light enters parallel to the absorbing layer, rather than perpendicular to it as was the case with the surface illuminated photodiode. However, to capitalize on this feature requires two additional design details. One is to make the absorbing region an optical waveguide. The other is to utilize traveling wave electrodes whose propagation velocity is matched to the propagation velocity of the optical waveguide. This combination largely avoids the bandwidth–responsivity tradeoff of surface illuminated photodiodes. The optical waveguide permits increasing the absorption volume, by increasing the wave- guide length, without the penalty of reduced bandwidth due to increased electrode capacitance – because of the traveling wave electrodes. The performance improve- ment of the traveling wave photodiode over the surface illuminated photodiode is impressive; Kato *et al.* (1992) reported a 50 GHz traveling wave photodiode with a responsivity of 0.85 A/W vs. a 30 GHz surface illuminated design with a responsivity of 0.92 A/W. Further, a responsivity greater than 0.5 A/W was demon- strated in a traveling wave photodiode with a bandwidth of 110 GHz (Kato *et al.*, 1994).

However, the traveling wave photodiode is not free of design tradeoffs. Prime among them is the fiber coupling efficiency. The refractive index step between the core and surround is larger in the photodiode optical guide than it is in an optical fiber. In turn this means that the fiber's optical mode diameter is larger than the photodiode's. The larger refractive index step also means higher scattering losses due to roughness at the core–surround interface. This loss, together with the residual velocity mismatch between the optical and electrode velocities, sets a practical limit on the maximum length of traveling wave photodiodes.

4.3 Passive impedance matching to modulation and photodetection devices

To explore the effects of RF filtering and impedance matching on intrinsic gain and frequency response, we need to add reactive components to the various device mod- els for the modulation and photodetection devices that were discussed in Chapter 2. Reasonably detailed, lumped-element, small-signal impedance models of all these devices that accurately characterize the device's measured impedance have been widely reported. However, the complexity of these models often obscures some key points. Hence for the purposes of this discussion we add one, and in a few cases two, reactive element(s) to each of the otherwise resistive impedance models of Chapter 2. Although at first this may seem to be an absurdly extreme simplifica- tion, it turns out to be sufficient for us to explore a number of important issues associated with gain, bandwidth and impedance matching.

Implicit with the mention of "lumped-element" above, is the restriction in the following discussion to frequencies and bandwidths where such element models are valid; i.e. for the most part we will not be considering frequencies where traveling wave effects need to be included.

One of the key ways to divide approaches to passive impedance matching is by whether they are lossy or lossless. It might seem that all physically realized approaches are lossy, since all actual components have finite Q and hence some amount of loss. However, what is generally meant by these categories is whether any *intentional* loss, in the form of one or more resistors, is included in the matching circuit.

Another key way of categorizing impedance matching is via the matching criterion. It can be shown from elementary network theory (see for example Van Valkenburg, 1964) that the maximum power transfer between a source and a load occurs when the load impedance is the complex conjugate of the source impedance. If we represent impedance as $Z = R + jX$, then the conjugate match condition can be written as

$$Z_S = Z_{LOAD}^* \Leftrightarrow R_S = R_{LOAD} \quad \text{and} \quad X_S = -X_{LOAD}; \tag{4.9}$$

i.e. a conjugate match is achieved when the real parts of the source and load are equal and the imaginary parts have opposite sign. For example, if the load impedance is a series RC, then the conjugate match is a series RL. This example suggests a fact that is generally true about conjugate matches: they are tuned or narrowband matches. The complexity of a conjugate match increases considerably as one attempts to maintain a conjugate match over wider frequency ranges.

We have actually been using conjugate matches all along in Chapters 2 and 3, without specifically identifying those matches as such. However, in the case of the purely resistive loads considered in those chapters, there are no imaginary parts. Consequently the conjugate match simplifies to matching the load resistance to the source resistance, which is what we did.

When the load impedance is expanded to include reactive components – as we plan to do in this chapter – then strictly speaking a magnitude match is insufficient and a conjugate match is required. However, if the match is only needed over a band of frequencies where there is negligible contribution from the reactive components, then a magnitude match can – and often is – used. For example, including the capacitance in the photodiode impedance results in a low pass frequency response. If the match is only needed for frequencies less than the pass band 3-dB frequency, then a magnitude match is sufficient. The condition for a magnitude match can be stated as

$$\text{Re}(Z_L) = R_S. \tag{4.10}$$

In contrast to the conjugate match, the magnitude match is usually simpler to implement and easier to realize effectively over a broad bandwidth. However, the magnitude match is limited to a specific class of load impedances.

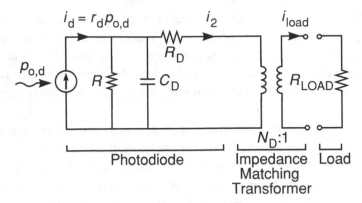

Figure 4.11 Schematic of a simple, low frequency photodiode circuit model connected to an ideal transformer for magnitude matching the photodiode to a resistive load.

Finally, in the sections below, we reverse the presentation order we have used previously and present the photodetector case first. This order permits us to show link results immediately following the discussion of each modulation device.

4.3.1 PIN photodiode

4.3.1.1 Broad bandwidth impedance matching

To explore the effects of gain and frequency response at the photodiode, we add only a single reactive element – a capacitor, C_D – to the basic current source photodiode model we introduced in Chapter 2. We also need to add two resistances; one in parallel with the current source, R, to account for the fact that the current source is not ideal and one in series with the current source, R_D, to represent the finite contact resistance.

Lossless, *magnitude* impedance matching can be implemented by connecting an ideal transformer between the photodiode and the load. To extract the maximum power from a current source, we want to make the load resistance as large as possible. The transformer accomplishes this by magnifying the load resistance as viewed from the photodiode current source. For the purposes of this development, we assume that the transformer is ideal in the sense that it is lossless and has negligible self and leakage inductances. Such a transformer can be completely characterized by its turns ratio, N_D. The resulting lumped-element, small-signal model of the photodiode is shown in Fig. 4.11. This circuit can also be used to represent the unmatched case by letting $N_D = 1$.

Lossy impedance matching to a photodiode is rarely used. To appreciate the tradeoffs this imposes, consider the two ways one might attempt to implement such a match. One way we might try is to put a resistor in series with R_D. However, recall that a photodiode is best modeled as a current source, whose impedance in the ideal case is infinite.[4] Consequently adding a matching resistor in series with R_D has no effect on the photodiode impedance.

The other way to try lossy impedance matching would be by placing a matching resistor, R_{MATCH}, in parallel with the photodiode capacitance, i.e. in parallel with resistor R in Fig. 4.11. The matching resistor value is then chosen such that $(R_{MATCH}\|R) + R_D = R_{LOAD}$, where the pair of vertical parallel bars represent the parallel connection of the two components.[5] However, adding this matching resistor would decrease the power delivered to the load, by as much as 3 dB in the limiting case where $R_D = 0$. On the other hand, omitting this resistor means that there is an impedance mismatch between the photodiode and the load, which in turn leaves one open to the deleterious effects of reflected RF power. In practice, the effects of reflected power can be minimized by keeping the distance between the photodiode and the load as short as possible.

To begin the derivation of the incremental detection efficiency for the transformer-matched, frequency-dependent photodiode, we start with an expression for the current flowing through the photodiode side of the transformer, i_2, as a function of the photodiode current, i_d:

$$i_2 = \frac{[R\|(1/sC_D)]}{[R\|(1/sC_D)] + R_D + N_D^2 R_{LOAD}} i_d. \tag{4.11}$$

Ideally the incremental model of a photodiode is a current source, in which case $R \to \infty$. For real photodiodes, R is obviously finite, but typically $R \gg R_{LOAD}$. Therefore to maximize the signal power that we can obtain from the photodiode, we want to present as large an impedance as possible to the photodiode. Thus the photodiode-to-load impedance matching is accomplished via a step-down transformer with the step-down going from photodiode to load. Consequently the current through the load in terms of i_2 is simply $i_{load} = N_D i_2$. The power supplied to the load, p_{load}, is just $i_{load}^2 R_{LOAD}$. Sequential substitutions of the above relations followed by solving for $p_{load}/p_{o,d}^2$ yields the desired result:

$$\frac{p_{load}}{p_{o,d}^2} = \frac{r_d^2 N_D^2 R_{LOAD}}{\left(sC_D(R_D + N_D^2 R_{LOAD}) + 1\right)^2}, \tag{4.12}$$

where we have neglected R, the resistance in parallel with the photodiode capacitance, since it is typically so large as to not affect the frequency response significantly.

---

[4] One of the factors that can violate this ideal is operation of the photodiode at high optical power. The large resulting photocurrent can result in a significant voltage drop across R_D that in turn decreases the reverse bias across the PIN junction. The reduced bias changes the junction capacitance, which alters the impedance.

[5] The "$\|$" symbol will be used to denote the parallel connection of two impedances, Z_1 and Z_2: $Z_1\|Z_2 = Z_1 Z_2/(Z_1 + Z_2)$.

One may well ask: without R, what is the basis for selecting the value of the matching transformer turns ratio in the equation $R = N_D^2 R_{\text{LOAD}}$? The answer is that since R is so large, it is rarely feasible to select N_D on the basis of impedance matching to R, from both practical and bandwidth considerations. For example, matching a 50-ohm load to a typical value of $R = 100\,000\ \Omega$ would require a transformer with $N_D \cong 45$, which is well beyond the capability of present high frequency transformers. Even if one could obtain such a transformer, it is unlikely one would use it because of the severe bandwidth reduction it would impose, as explained further in conjunction with (4.14) immediately below. However, consistent with the bandwidth requirements, we do want to select the turns ratio as large as possible to maximize the signal power as discussed following (4.11).

We now wish to explore analytically this gain vs. bandwidth tradeoff as a function of the matching transformer turns ratio. The case without impedance matching can be obtained from (4.12) by setting the turns ratio to 1. We first calculate the low frequency gain enhancement with impedance matching relative to the gain without impedance matching by setting $s = 0$ in (4.12); the result is

$$\text{Gain increase} = \frac{(4.12)|_{N_D}}{(4.12)|_{N_D=1}} = N_D^2. \tag{4.13}$$

Next we calculate the bandwidth with impedance matching relative to the bandwidth without impedance matching by taking the ratio of the coefficients of s in (4.12) when $N_D > 1$ to the case when $N_D = 1$:

$$\text{Bandwidth increase} = \frac{R_D + R_{\text{LOAD}}}{R_D + N_D^2 R_{\text{LOAD}}} \cong \frac{1}{N_D^2} \bigg|_{\substack{N_D \text{ large or} \\ R_{\text{LOAD}} > R_D}}, \tag{4.14}$$

where in calculating (4.14) it is important to keep in mind that the bandwidth in each case is proportional to the reciprocal of the coefficients of s in (4.12). Since $N_D^2 > 1$, the gain increases, but the bandwidth decreases.

As an example of (4.13) and (4.14) consider the matched case where $N_D^2 = 10$. Further assume that $R_D = 5\ \Omega$ and $R_{\text{LOAD}} = 50\ \Omega$. The gain increase is 10 or 10 dB over the unmatched case and the bandwidth increase is 0.12, which is actually a decrease. In the limiting case of (4.14) the bandwidth decreases with increasing matching transformer turns ratio at the *same* rate that the gain increases, (4.13). Therefore the gain–bandwidth product remains constant in this limit. Thus the turns ratio presents the link designer with a classic gain–bandwidth tradeoff: passive impedance matching permits increased intrinsic gain, but only at the expense of reduced bandwidth.

To see graphically the effects of the matching transformer turns ratio on gain and bandwidth, we need to assume a value for the photodiode responsivity and

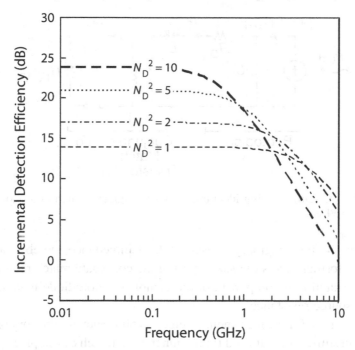

Figure 4.12 Calculated photodiode incremental detection efficiency vs. frequency with matching transformer turns ratio as the parameter for the circuit shown in Fig. 4.11.

capacitance; we choose 0.7 A/W and 0.5 pF, respectively as representative of present photodiodes. Figure 4.12 is a plot of (4.12) using the values assumed above and four transformer turns-squared ratios: 1, which is the unmatched case, 2, 5 and 10. Figure 4.12 makes the gain–bandwidth tradeoff clear: the higher the gain, the lower the bandwidth.

It is important to note that these changes are due to the matching *method*, not the *degree* of match, since in both cases a perfect match was assumed. The effects of impedance mismatch are discussed in Section 4.4.

4.3.1.2 Bandpass impedance matching to a photodiode

We now consider a lossless *conjugate* match between the photodiode and the load. By its very nature, such a match is achieved over only a limited range of frequencies and consequently this form of match imposes a bandpass frequency response. To achieve a conjugate match to the photodiode capacitance simply requires adding an inductance, L, in series with the photodiode resistance, R_D, as shown in Fig. 4.13. In this case the inductive reactance meets the *conjugate* part of the conjugate match requirement by supplying an imaginary impedance with the

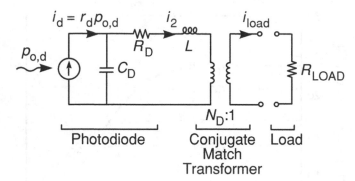

Figure 4.13 Circuit showing ideal transformer conjugate match of a photodiode to a resistive load.

opposite sign of the imaginary capacitive load impedance thereby canceling it out. The transformer meets the *match* part of the conjugate match requirement by setting the residual real component of the combined photodiode-matching circuit equal to the source resistance.

The development for the incremental detection efficiency for the conjugate match follows a path similar to that used for the magnitude match development presented above. The relationship between i_2 and i_d for the conjugate match, which is analogous to (4.11) in the magnitude match, is

$$i_2 = \frac{(1/sC_D)}{(1/sC_D) + R_D + sL + N_D^2 R_{\text{LOAD}}} i_d, \qquad (4.15)$$

where we have neglected the photodiode shunt resistance, R, at the outset this time. Subjecting (4.15) to the same series of sequential substitutions that we did in going from (4.11) to (4.12) above, we arrive at the expression for the incremental detection efficiency of a conjugately matched photodiode:

$$\frac{p_{\text{load}}}{p_{o,d}^2} = \frac{r_d^2 N_D^2 R_{\text{LOAD}}}{\left(s^2 C_D L + s C_D\left(R_D + N_D^2 R_{\text{LOAD}}\right) + 1\right)^2}. \qquad (4.16)$$

The bandpass frequency response with the conjugate match is evident from the second-order denominator of (4.16), which is then squared to obtain a fourth-order response. As a consistency check, we note that without the inductance the conjugate match should reduce to the magnitude match, which is what (4.16) does when $L = 0$.

To help visualize the changes in frequency response as a function of matching transformer turns ratio in the conjugate match case, we plot (4.16) in Fig. 4.14 using the same values and for the same range of N_D as we did in Fig. 4.12. The value of the series inductance was fixed for these calculations at 25 nH.

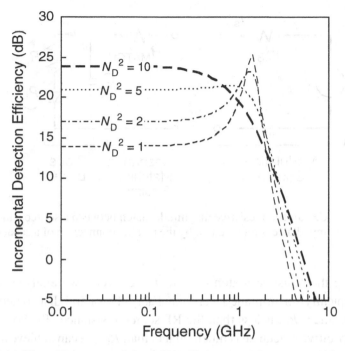

Figure 4.14 Plot of incremental detection efficiency vs. frequency with matching transformer turns ratio as the parameter for the circuit shown in Fig. 4.13.

We see a range of frequency response shapes from peaking for the lowest transformer turns ratio-squared (underdamped) to roll-off at the highest transformer turns-squared ratio (overdamped). However, for $N_D^2 = 5$ the peaking just balances the roll-off (critically damped) thereby maximizing the bandwidth. This suggests that in the conjugate match case, inductance as well as the turns ratio needs to be selected for optimum performance.

We now have the photodiode impedance-matched frequency responses that we need for the link calculations we wish to do in conjunction with the modulation device sections that follow.

4.3.2 Diode laser

4.3.2.1 Broad bandwidth impedance matching

For our investigation of passive matching to a diode laser, we choose two *magnitude* matching approaches: a single series resistor for the lossy case and an ideal transformer for the lossless case. In the next section we explore a simple lossless *conjugate* match. For all these cases we assume that the frequency response of the matching circuit rolls off well before the laser relaxation resonance.

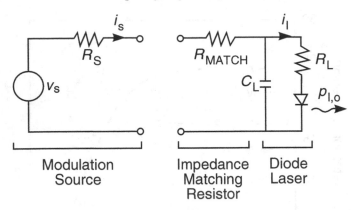

Figure 4.15 Circuit for a resistive magnitude match between a source and a diode laser whose impedance is represented by the parallel connection of a capacitor and a resistor.

We develop the resistive match case first; the circuit we analyze is shown in Fig. 4.15. Typically for in-plane lasers (i.e. Fabry–Perot and distributed feedback) the laser resistance, R_L, is less than the RF source resistance, R_S. For frequencies where the capacitive reactance is much greater than R_L, we can achieve a magnitude impedance match between the laser and the source by inserting a series resistor, R_{MATCH}, whose value is selected such that the following equality holds: $R_{MATCH} + R_L = R_S$.

Our basis for comparing the effectiveness of the various matching methods is the incremental modulation efficiency. To obtain this quantity for the resistive match case, we begin as we did when developing the modulation efficiency for the purely resistive laser in Section 2.2.1.1, (2.8), with the expression that relates the laser optical power to the laser current:

$$p_{l,o} = s_l i_l. \tag{4.17}$$

With a capacitor, C_L, in parallel with the laser resistance, the laser current is no longer equal to the source current, as it was in the purely resistive case. Straightforward circuit analysis can be used to determine the fraction of the source current that flows through the laser resistance to be

$$i_l = \left(\frac{1}{s\tau + 1} \right) i_s, \tag{4.18}$$

where $\tau = C_L R_L$. The current drawn from the source is simply the ratio of the source voltage to the total impedance of the loop:

$$i_s = \frac{v_s}{R_S + R_{MATCH} + [(1/sC_L) \| R_L]}. \tag{4.19}$$

We can now solve for the modulated laser power in terms of the source voltage by substituting (4.19) into (4.18), then substituting the result into (4.17). Some algebra is required on the resulting expression to keep the frequency dependent term in the normalized form $s\tau + 1$. The result of all these steps is the following expression for the modulated optical power as a function of the source voltage and the circuit impedances:

$$p_{\ell,0} = \frac{s_\ell v_{\mathrm{s}}}{R_{\mathrm{S}} + R_{\mathrm{MATCH}} + R_{\mathrm{L}}} \left(\frac{1}{\dfrac{s C_{\mathrm{L}} R_{\mathrm{L}} (R_{\mathrm{S}} + R_{\mathrm{MATCH}})}{R_{\mathrm{S}} + R_{\mathrm{MATCH}} + R_{\mathrm{L}}} + 1} \right). \qquad (4.20)$$

To obtain the incremental modulation efficiency, we need to divide the square of (4.20) by the available power from the source, $v_{\mathrm{s}}^2/4R_{\mathrm{S}}$. Although this may appear to be a daunting task at first glance, it is actually quite simple if we realize two facts. One is that the division only affects the term preceding the parentheses; the other is that under the match condition, the denominator of this term is equal to $2R_{\mathrm{S}}$. With these facts in mind, the incremental modulation efficiency for a resistive match to a parallel RC laser model is

$$\left. \frac{p_{\ell,0}^2}{p_{\mathrm{s,a}}} \right|_{\substack{\text{resistive} \\ \text{match}}} = \frac{s_\ell^2}{R_{\mathrm{S}}} \left(\frac{1}{\dfrac{s C_{\mathrm{L}} R_{\mathrm{L}} (R_{\mathrm{S}} + R_{\mathrm{MATCH}})}{2R_{\mathrm{S}}} + 1} \right)^2. \qquad (4.21)$$

As a check on the validity of (4.21) we note that its value at dc, i.e. for $s = 0$, is the same as the incremental modulation efficiency for the diode laser model without the parallel capacitance, (2.11). Thus (4.21) shows that including a capacitor across the laser resistance does not affect the dc value, but simply limits the bandwidth over which this dc value is maintained: the laser frequency response now has a low pass form. We postpone further discussion of the resistive match case until we have developed the corresponding expression for the transformer match case.

For the transformer match case we again assume that the transformer is ideal and consequently that it can be completely characterized by its turns ratio, N_{L}. Figure 4.16 shows a schematic of an ideal transformer match between a source and the same diode laser impedance model we just used for the resistive match. As in the resistive match case, we want to achieve a magnitude match between the source resistance, R_{S}, and the laser resistance, R_{L}. To achieve the desired match with a transformer, its turns ratio must be chosen such that

$$N_{\mathrm{L}}^2 R_{\mathrm{L}} = R_{\mathrm{S}}. \qquad (4.22)$$

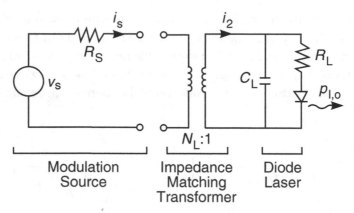

Figure 4.16 Circuit for an ideal transformer magnitude match between a source and a diode laser whose impedance is represented by the parallel connection of a capacitor and a resistor.

The derivation of the incremental modulation efficiency for the transformer match case can begin with (4.17), since we are using the same laser model we used for the resistive match case. We can also use (4.18) if we substitute the secondary current, i_2, for the source current. For an ideal transformer, the secondary current is related to the primary current by the turns ratio, viz.:

$$i_2 = N_L i_s, \tag{4.23}$$

where in this simple circuit the primary current is the source current.

For the source current expression, we need to determine the total impedance seen by the source. This is simply the laser impedance, magnified by the transformer turns ratio squared, in series with the source resistance. Therefore the expression for the source current is

$$i_s = \frac{v_s}{R_S + N_L^2[(1/sC_L) \| R_L]}. \tag{4.24}$$

Time for a little more algebra: expand out the parallel impedance in (4.24), substitute (4.24) into (4.23), put the result into (4.18) – with the substitution of i_2, for source current – and finally substitute all this into (4.17). The result is

$$p_{\ell,o} = \frac{s_\ell N_L v_s}{R_S + N_L^2 R_L} \left(\frac{1}{\dfrac{sC_L R_S R_L}{R_S + N_L^2 R_L} + 1} \right). \tag{4.25}$$

Applying (4.22) to the two resistive denominator terms in (4.25) simplifies each of these terms to $2R_S$. Squaring (4.25) and dividing by the available power from the

source yields the expression for the incremental modulation efficiency of a diode
laser with an ideal transformer match:

$$\frac{p_{l,o}^2}{p_{s,a}}\bigg|_{\substack{\text{transformer}\\\text{match}}} = \frac{s_l^2 N_L^2}{R_S}\left(\frac{1}{\dfrac{sC_L R_L}{2}+1}\right)^2.\tag{4.26}$$

Equation (4.26) has the same general form as (4.21) but with two important
differences in the details. We explore these differences with the aid of Fig. 4.17.
Since we wish to focus on the effects of various laser resistances, we assume that
the source resistance is fixed at 50 Ω. Further we assume throughout this section
that the laser slope efficiency is 0.045 W/A – a typical value for early Fabry–Perot
lasers – and the laser capacitance is 20 pF. With these assumed values, Fig. 4.17(a)
is a plot of (4.21) – the resistive match case – for three values of R_L. As we can
see, the dc and low frequency values of the incremental modulation efficiency are
independent of the value of R_L, since the matching condition has ensured that there
is no change in the current through R_L. The bandwidth, however, does increase as
the laser resistance decreases.

Also included in Fig. 4.17(a) is the unmatched case, i.e. when the matching resistor, R_{MATCH}, is set to zero. When viewed from the perspective of intrinsic gain,
this case does offer improved gain over the resistive match case. However, as we
shall see in Chapter 5, the intrinsic gain increase achieved via omitting the matching
resistor does not result in a corresponding reduction in the noise figure.

Figure 4.17(b) is the analogous plot for the case of a transformer match – as
described by (4.26) – using the same values for R_L and C_L. The bandwidth is again
an inverse function of R_L. However, the dc and low frequency responses are also
inversely related to R_L, in contradistinction to the resistive match case.

For the diode laser we can also compare the gain and bandwidth for the resistive
and transformer matching cases. The increase in dc and low frequency gain in the
transformer match case compared to the resistive match case is simply the ratio of
the terms in front of the parentheses in (4.26) to (4.21); the result is

$$\text{Gain increase} = N_L^2 = \frac{R_S}{R_L},\tag{4.27}$$

which indicates that as the laser resistance decreases, the gain increases without
limit. This may be a surprising result for those who that recall that under the
resistive match case, the maximum increase in link gain with $R_L = R_{MATCH} = 0$ was
6 dB. However, the apparent conflict between these two cases is resolved when we

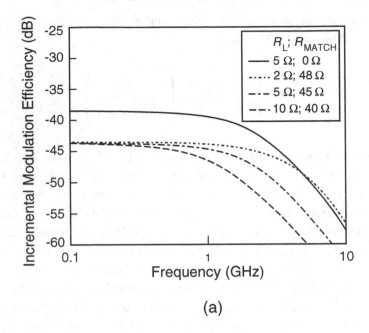

(a)

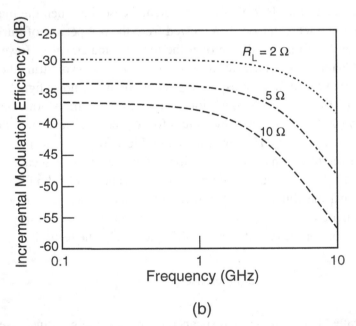

(b)

Figure 4.17 Plot of the incremental modulation efficiency vs. frequency for (a) resistive and (b) transformer match with laser resistance as the parameter.

consider that a transformer – at least an ideal one – has the potential to increase the current by more than a factor of 2, which is the maximum current increase in the case of resistive matching.

The bandwidth increase in going from resistive to transformer matching can be calculated from the inverse of the ratio of the coefficients of s in (4.26) to (4.21). This ratio turns out to be

$$\text{Bandwidth increase} = \frac{R_S + R_{MATCH}}{R_S} = 2 - \frac{R_L}{R_S}, \qquad (4.28)$$

where we have used the match condition, $R_{MATCH} + R_L = R_S$, to write the second form of (4.28). Note that this form makes it clear that as the laser resistance decreases, the maximum bandwidth increase is limited to a factor of 2.

To get a feel for the magnitude of these changes, assume that $R_L = 5\ \Omega$. Then for a 50 Ω source resistance, the gain increase is a factor of 10, or 10 dB, while the bandwidth increase is a factor of 1.9 or about 3 dB. In contrast to the photodiode case, the diode laser transformer match increases the gain–bandwidth product since both the gain *and* the bandwidth increase as the turns ratio increases. This is an unusual result. The basis for it lies in the fact that the gain increases as the laser resistance decreases, which is counter to the trend for many devices, such as the photodiode discussed above and the modulator to be discussed below.

One question that arises at this point is: given the improvements transformer matching offers, why would anyone use resistive matching? The answers are generally cost and realizability. Resistors are less expensive than transformers. As for realizability, actual resistors much more closely approximate an ideal resistor than actual transformers approximate an ideal transformer. For example, we have assumed that the loss of the transformer can be neglected, a condition that often is not valid in practice. From a frequency response perspective, a resistive match extends down to dc whereas a transformer, which can have a bandwidth that extends over several orders of magnitude, cannot extend down to dc. Thus, as with many other design choices, the selection between resistive and transformer match involves tradeoffs among several parameters.

Equations (4.27) and (4.28) quantify the tradeoff between gain and bandwidth between the two matching methods. However, these equations also point out that the gain increases more than the bandwidth decreases, so there is a net *increase* in the gain–bandwidth product in going from a resistive to a transformer match. Thus, another question that arises at this point is: given the increase in gain–bandwidth product we have seen in going from a resistive to a transformer match, what additional improvements are possible using other matching methods? The answer is that slightly more improvement is possible, but to substantiate this answer we need the Bode–Fano limit – which we present in Section 4.4. However, we can use (4.21) and (4.26) to anticipate the form of this limit.

The bandwidth in (4.21) and (4.26) is proportional to the reciprocal of the coefficient of s. Thus the bandwidth in the resistive match case is proportional to

$$\text{Resistive match bandwidth} \propto \frac{2}{C_L R_L} \left(\frac{R_S}{R_S + R_{MATCH}} \right), \qquad (4.29)$$

whereas the bandwidth of the transformer match is proportional to

$$\text{Transformer match bandwidth} \propto \frac{2}{C_L R_L}. \qquad (4.30)$$

Thus the bandwidth improvement that transformer matching offers over resistive matching comes from the transformer's ability to eliminate the second term in (4.29). But in both cases the bandwidth is basically determined by a constant divided by an RC product. Consequently the question as to the maximum bandwidth improvement that some hypothetical matching method could achieve can be rephrased as: what is the maximum constant in the numerator of expressions like (4.30)? We explore the answer to this question in Section 4.4.

4.3.2.2 Bandpass impedance match to a diode laser

We now explore the frequency response of the incremental modulation efficiency for the simplified reactive impedance model of a diode laser with a conjugate match to a resistive source. Since, like the photodiode, the laser's dominant reactive component is a capacitor,[6] this type of match can easily be achieved by inserting an inductor in series between the diode laser and the transformer. The resulting circuit is shown in Fig. 4.18.

The derivation for the conjugate match basically follows the derivation of the magnitude match using a transformer, starting with (4.22). The transformer turns ratio is changed only to accommodate the additional series resistor, R_{ML}:

$$N_L^2(R_{ML} + R_L) = R_S, \qquad (4.31)$$

where R_{ML} represents the loss in the inductor L.

The relationship between primary and secondary current, (4.23), remains the same. The expression for the total current drawn from the source, (4.24) also needs to be modified by adding R_{ML} and L in series with the parallel $R_L C_L$ impedance:

$$i_s = \frac{v_s}{R_S + N_L^2[R_{ML} + sL + (1/sC_L)\|R_L]}. \qquad (4.32)$$

---

[6] Also like the photodiode, the laser capacitance is not fixed but is a weak function of bias. For the purposes of this first-order development we neglect this dependence.

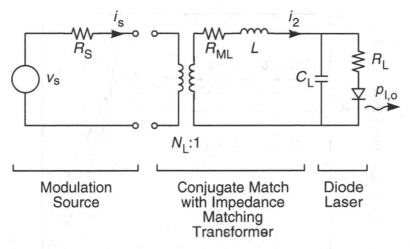

Figure 4.18 Schematic for an ideal transformer conjugate match between a source and a diode laser whose impedance is represented by the parallel connection of a capacitor and a resistor.

Once again it's time for more algebra: expand out the denominator of (4.32), substitute the result into (4.23), put that result into (4.18) and finally substitute all this into (4.17):

$$p_{l,o} = \frac{s_t N_L v_s}{2R_S \left(s^2 \dfrac{C_L R_L L}{2(R_{ML} + R_L)} + s \left(\dfrac{L + C_L R_L R_{ML}}{2(R_{ML} + R_L)} + \dfrac{C_L R_L}{2} \right) + 1 \right)}, \quad (4.33)$$

where we have made repeated use of (4.31) to get the form given in (4.33).

To obtain an expression for the incremental modulation efficiency of a directly modulated diode laser with a series resonant conjugate impedance match simply requires squaring (4.33) and dividing by the available power:

$$\left. \frac{p_{l,o}^2}{p_{s,a}} \right|_{\substack{\text{conjugate} \\ \text{match}}} = \frac{s_t^2 N_L^2}{R_S} \left(\frac{1}{s^2 \dfrac{C_L R_L L}{2(R_{ML} + R_L)} + s \left(\dfrac{L + C_L R_L R_{ML}}{2(R_{ML} + R_L)} + \dfrac{C_L R_L}{2} \right) + 1} \right)^2. \quad (4.34)$$

As a sanity check on (4.34), we note that by setting both of the just introduced components to zero, the above expression reduces to (4.26).

Although the dc value of (4.34) is the same as the dc value of (4.26), the bandpass shape of the conjugate match is evident from the second-order resonant term in the denominator of (4.34). When R_{ML} is set to zero in (4.34), the peak value of (4.34)

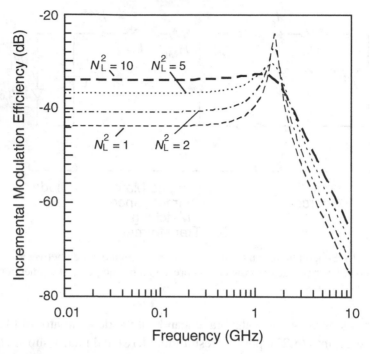

Figure 4.19 Plot of incremental modulation efficiency vs. frequency with matching transformer turns ratio as the parameter for the circuit shown in Fig. 4.18.

can be higher than the peak value of (4.26) since there are frequencies for which the denominator of (4.34) is less than the denominator of (4.26).

To get a feel for the gain vs. bandwidth tradeoffs introduced by transformer conjugate matching, we plot (4.34) in Fig. 4.19 for the same laser parameters and the same range of N_L^2 that we used in Fig. 4.17. The value of R_{ML} is chosen subject to the constraint $N_L^2(R_{ML} + R_L) = R_S$.

4.3.2.3 Direct modulation link with passive impedance matching

We now have most of the material necessary to construct a model for the frequency response of a complete, passively impedance-matched link. The one remaining piece we need is the expression for the frequency response of the *link* in terms of the frequency responses of the impedance-matched modulation and photodetection *devices*. In general this is a hard problem. As we have seen from the circuit analyses in the previous sections, the frequency response of a cascade connection of electrical components is not simply the product of the frequency responses of each of the components taken individually.

However, in the case of an optical link, the link frequency response *is* simply the product of the frequency responses of the impedance-matched modulation and photodetection circuits! The reason that an optical link analysis can avoid the complexities encountered with cascading electrical components is the isolation from circuit loading provided by the electro-optic devices. Changing the photodiode matching circuit does not change the load impedance seen by the modulation device matching circuit, and vice versa.[7] This important result greatly simplifies the frequency response analysis of optical links.

There have been many applications that use the impedance matching techniques presented in Sections 4.3.1 and 4.3.2 to increase the link gain. We focus here on demonstrations of lossless,[8] bandpass impedance matching, several of which have been reported in the literature (Ackerman *et al.*, 1990; Goldsmith and Kanack, 1993; Onnegren and Alping, 1995). Because of their modeling and measurements of both gain and noise figure, which is discussed in the next chapter, we choose the work of Onnegren and Alping to illustrate what can be achieved in practice and to compare these results with the predicted values.

In the frequency range of their link, around 4 GHz, they used a cascade of transmission lines – each with different dimensions – to implement the transformer matching function. Unfortunately the frequency response of transmission line transformers is not independent of frequency like the "ideal" transformers we have assumed in the analysis so far. The original work of Onnegren and Alping modeled this frequency dependence by adding lumped elements to the impedance models for the laser and photodetector we have used here. Although they achieved excellent agreement between their modeled and measured results, the resulting equations are so complex that they obscure our goal here: to present the basic principles and techniques as clearly as possible. Consequently we neglect the frequency dependence of the transmission line – and the additional lumped elements required to model it. Thus we expect the measured response to exhibit a roll-off both above and below the matching band, which our simplified model response will not show. Once we understand the techniques, an arbitrary degree of accuracy can be obtained by including an increasing number of elements in the model.

Figure 4.20 shows our simplified version of the Onnegren and Alping models for a diode laser and a PIN photodiode. By proper choice of dimensions, the transmission line transformers can also have an inductive output impedance. Consequently there is no explicit series inductor shown in Fig. 4.20, as there was in the development of the bandpass match.

For each device, we wish to consider the frequency response with and without lossless, bandpass impedance matching. To calculate the expected frequency

---

[7] Technically this result follows from the fact that a link is non-reciprocal; i.e. the response at one terminal pair (e.g. the link output) due to the stimulation at another terminal pair (e.g. the input) is not the same as the opposite stimulus–response condition.

[8] Of course since we are now discussing actual – as opposed to ideal – matching, these are demonstrations of *near*-lossless matching.

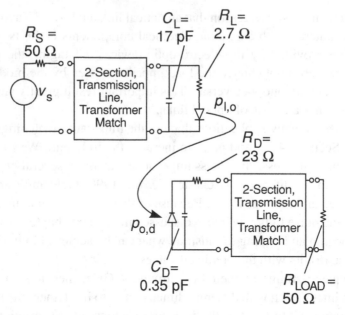

Figure 4.20 Schematic of simplified circuit for bandpass matching to a diode laser and a PIN photodetector (based on Onnegren and Alping, 1995).

response for each of the four cases, we use the link frequency response decomposition property stated above and simply multiply the frequency responses of the impedance-matched laser and photodetector. Taking the appropriate pairwise combinations of either (4.21) with $R_{\text{MATCH}} = 0$ or (4.34) and combining them with either (4.12) with $N_{\text{D}} = 1$ or (4.16) we immediately obtain the link frequency response for each case:

$$g_{i,\text{uml,umd}} = \frac{s_\ell^2 r_{\text{d}}^2}{\left(\dfrac{sC_{\text{L}}R_{\text{L}}}{2} + 1\right)^2 (sC_{\text{D}}(R_{\text{D}} + R_{\text{S}}) + 1)^2}; \tag{4.35}$$

$$g_{i,\text{ml,umd}} =$$

$$\frac{s_\ell^2 r_{\text{d}}^2 N_{\text{L}}^2}{\left(s^2 \dfrac{C_{\text{L}}R_{\text{L}}L_{\text{L}}}{2(R_{\text{M}} + R_{\text{L}})} + s\left(\dfrac{L_{\text{L}} + C_{\text{L}}R_{\text{L}}R_{\text{M}}}{2(R_{\text{M}} + R_{\text{L}})} + \dfrac{C_{\text{L}}R_{\text{L}}}{2}\right) + 1\right)^2 (sC_{\text{D}}(R_{\text{D}} + R_{\text{S}}) + 1)^2};$$

$$\tag{4.36}$$

$$g_{i,\text{uml,md}} = \frac{s_\ell^2 r_{\text{d}}^2 N_{\text{D}}^2}{\left(\dfrac{sC_{\text{L}}R_{\text{L}}}{2} + 1\right)^2 \left(s^2 C_{\text{D}}L_{\text{D}} + sC_{\text{D}}(R_{\text{D}} + N_{\text{D}}^2 R_{\text{S}}) + 1\right)^2}; \tag{4.37}$$

$$g_{i,ml,md} =$$

$$\frac{s_i^2 r_d^2 N_L^2 N_D^2}{\left(s^2 \frac{C_L R_L L_L}{2(R_M + R_L)} + s\left(\frac{L_L + C_L R_L R_M}{2(R_M + R_L)} + \frac{C_L R_L}{2}\right) + 1\right)^2 \left(s^2 C_D L_D + s C_D (R_D + N_D^2 R_S) + 1\right)^2},$$

(4.38)

where we have made use of the assumption that $R_S = R_{LOAD}$ to cancel these terms when combining the diode laser and photodiode expressions. In equations (4.35) through (4.38) the subscript "m" refers to the matched case while the subscript "um" refers to the unmatched case.

Figure 4.21(a) through (d) show the modeled frequency responses calculated using the equations just presented together with the measured gains for the four cases. The full frequency response for the case with unmatched laser and photodiode – it has a low frequency intrinsic gain of −42 dB and a 3-dB bandwidth of 3 GHz – is not shown, so that we can better display the frequency response of the bandpass cases. The unmatched passband extended from 2.3 GHz to 5.3 GHz, which is in contrast to (4.35), which predicts a passband that extends down to dc. The effects of neglecting the transmission line transformers' frequency dependence is particularly noticeable in the difference between the modeled and measured responses shown in Fig. 4.21(a).

As can be seen from these figures, matching at both the laser and photodiode reduces the bandwidth to 410 MHz, from 3 GHz in the unmatched case, while increasing the intrinsic gain to −22 dB, up from −42 dB in the unmatched case. Also as expected, the gain increase with matching at both laser and detector is simply the sum of the gain increases with the laser and detector matched individually. Although the agreement between model and measurement using the simplified model is reasonably good, the authors were able to achieve even better agreement by using additional elements in the impedance models for the laser and photodiode.

4.3.3 Mach–Zehnder modulator

4.3.3.1 Broad bandwidth impedance matching

In our initial discussion of a Mach–Zehnder modulator in Chapter 2, we mentioned – but quickly eliminated – the electrode capacitance from the modulator circuit model. We now need to restore this capacitance for the purposes of this chapter. But the resulting parallel *RC* circuit is the same as the one we already extensively analyzed for the diode laser!

However, there are two subtle differences between these two cases that prevent us from writing the required expressions by inspection. In the direct modulation of the diode laser, the modulated optical power was proportional to the *current*

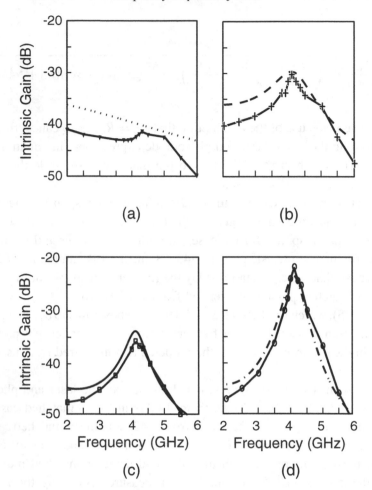

Figure 4.21 Modeled and measured link gain vs. frequency with (a) no impedance matching at either laser or detector, (b) laser impedance matching only, (c) detector impedance matching only and (d) matching at both the laser and detector. (Data from Onnegren and Alping, 1995, reprinted with permission of the authors.)

through the laser resistance, whereas in the case of the Mach–Zehnder modulator, the modulated optical power is proportional to the *voltage across* the parallel resistance and capacitance. The other difference is that, unlike the diode laser, the modulator parallel resistance is typically much larger than the source resistance.[9] Consequently to implement a resistive magnitude match to a modulator requires an external parallel resistor, whose value is chosen such that when it is connected in parallel with the modulator resistance, the result is equal to the source resistance.

---

[9] There is also a resistance in series with this parallel *RC* combination that is typically much less than even the source resistance. When $R_M = 1000 \, \Omega$, this series resistance is about 5 Ω. Thus it is reasonable to neglect the series resistance in a first-order model such as we are proposing here.

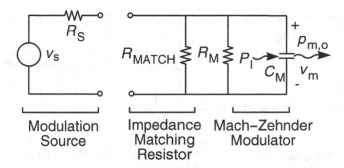

Modulation Source · Impedance Matching Resistor · Mach–Zehnder Modulator

Figure 4.22 Schematic of a resistive magnitude match between a resistive source and the parallel connection of two lumped elements – R and C – which represent a first-order, low-frequency impedance model for a Mach–Zehnder modulator.

The circuit for the resistive magnitude match between the modulation source and the frequency dependent Mach–Zehnder modulator is shown in Fig. 4.22.

We begin the derivation of the frequency dependent incremental modulation efficiency for the Mach–Zehnder modulator with a resistive magnitude match by recalling the small-signal transfer function, (2.14), that relates modulation voltage, v_m, to modulated optical power, $p_{m,o}$, for a Mach–Zehnder modulator that was developed in Section 2.2.2.1; it is repeated below for convenience:

$$p_{m,o} = \frac{T_{FF}P_I\pi}{2V_\pi}v_m. \tag{4.39}$$

To simplify the expressions below, we will represent the parallel combination of resistors R_{MATCH} and R_M by a single equivalent resistor R_E:

$$R_E = \frac{R_{MATCH}R_M}{R_{MATCH} + R_M}. \tag{4.40}$$

If we recognize that there is simply a voltage divider between the modulation voltage source, v_s, and the modulator voltage, we can write the required relationship between these two variables by inspection:

$$v_m = \frac{R_E||(1/sC_M)}{R_S + R_E||(1/sC_M)}v_s. \tag{4.41}$$

Expanding out the parallel RC combination in (4.41) and substituting the result into (4.39) yields

$$p_{m,o} = \left(\frac{T_{FF}P_I\pi}{2V_\pi}\right)\left(\frac{R_E}{R_S + R_E}\right)\frac{1}{\left(sC_M\dfrac{R_SR_E}{R_S + R_E} + 1\right)}v_s. \tag{4.42}$$

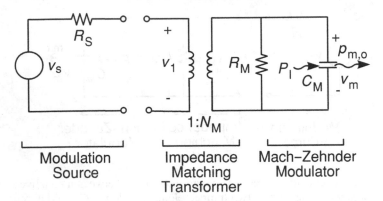

Figure 4.23 Schematic for an ideal transformer magnitude match between a source and the *RC* impedance model for a Mach–Zehnder modulator.

Assuming a match, i.e. $R_E = R_S$, and dividing the square of (4.42) by the available power of the modulation source yields the desired result:

$$\left.\frac{p_{m,o}^2}{p_{s,a}}\right|_{\substack{\text{resistive} \\ \text{match}}} = \left(\frac{T_{FF}\, P_I \pi}{2V_\pi}\right)^2 R_S \frac{1}{\left(\dfrac{s C_M R_S}{2} + 1\right)^2} = \frac{s_{mz}^2}{R_S} \frac{1}{\left(\dfrac{s C_M R_S}{2} + 1\right)^2},$$

(4.43)

where to obtain the second form of the incremental modulation efficiency we have again used the slope efficiency definition for a Mach–Zehnder we introduced in Section 2.2.2.1, viz.:

$$s_{mz} = \frac{T_{FF}\, P_I \pi\, R_S}{2V_\pi}.$$

(4.44)

By comparing the frequency dependent form of the incremental modulation efficiency, (4.43), with the dc form of the same parameter, (2.17), we see that the frequency dependent efficiency is simply the dc efficiency times a frequency dependent term. We also see from (4.43) that the efficiency is *independent* of the modulator resistance, which is the same result we observed with the diode laser under resistive match.

Next we consider replacing the lossy matching resistor with lossless magnitude matching via an ideal transformer, as shown in Fig. 4.23. Since $R_M > R_S$, we need a step-up transformer to establish the match condition: $N_M^2 R_S = R_M$.

The development for this case basically follows the same path we used for the resistive match case. The transformer establishes the relationship between the voltages at its terminals: $v_m = N_M v_1$. The relationship between v_1 and v_s can again be written by inspection:

$$v_1 = \frac{\dfrac{1}{N_M^2}[R_M||(1/sC_M)]}{R_S + \dfrac{1}{N_M^2}[R_M||(1/sC_M)]} v_s. \tag{4.45}$$

Expanding out the parallel RC combination in (4.45) and substituting the result into (4.39) yields

$$p_{m,o} = \left(\frac{T_{FF}P_I\pi}{2V_\pi}\right)\left(\frac{N_M^2 R_S}{N_M^2 R_S + R_M}\right)\frac{1}{\left(sC_M\dfrac{N_M^2 R_S R_M}{N_M^2 R_S + R_M} + 1\right)} v_s. \tag{4.46}$$

Assuming the transformer turns ratio is selected to establish a match and dividing (4.46) by the available power from the source yields the expression for the incremental modulation efficiency of a Mach–Zehnder modulator with a lossless, magnitude match:

$$\left.\frac{p_{m,o}^2}{p_{s,a}}\right|_{\substack{\text{transformer}\\\text{match}}} = \frac{s_{mz}^2 N_M^2}{R_S}\frac{1}{\left(\dfrac{sC_M R_S N_M^2}{2} + 1\right)^2}, \tag{4.47}$$

where we have used the matching condition $N_M^2 R_S = R_M$ to write both slope efficiency and the frequency response in the form shown in (4.47).

Figure 4.24 is a plot of (4.43) and (4.47) vs. frequency for the following typical set of modulator values: $P_I = 50$ mW, $T_{FF} = 3$ dB, $V_\pi = 5$ V, $C_M = 10$ pF and $R_S = 50\ \Omega$. Comparing the transformer match case, (4.47), to the resistive match case, (4.43), we see that the transformer match does produce a higher modulation efficiency – by a factor of N_M^2 – than the resistor match, which is in line with our intuition. This increase in slope efficiency is analogous to the increase we saw with the diode laser.

However, the bandwidth decrease for the Mach–Zehnder – unlike the diode laser cases – is also identically N_M^2, as can be easily demonstrated by taking the ratio of the coefficients of s in (4.47) to (4.43). Consequently the gain–bandwidth product of an impedance matched Mach–Zehnder modulator is a constant, independent of the matching transformer turns ratio. Thus the modulator, like the photodiode, is another example of a device that presents the link designer with a classic gain–bandwidth tradeoff. Of course the designer of an externally modulated link also has the option of changing the optical power, which permits control of the intrinsic gain that is independent of bandwidth.

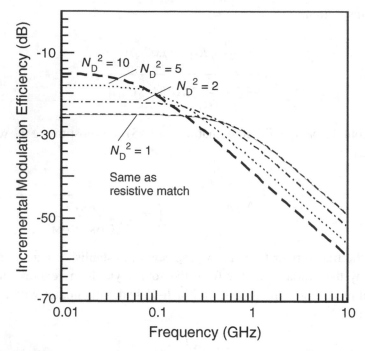

Figure 4.24 Modeled incremental modulation efficiency for (a) resistive match (4.43) and (b) transformer match (4.47) vs. frequency.

4.3.3.2 Bandpass impedance matching to a Mach–Zehnder modulator

The last impedance match case we explore is the conjugate match between a resistive source and the first-order RC impedance model we have been using for a low frequency Mach–Zehnder modulator. As we have with the previous conjugate match cases, the simplest way to implement such a match is to add a series inductor to the circuit shown in Fig. 4.23; the result is shown in Fig. 4.25. Although we show the inductor as a separate element, in practice it is often part of the circuit model for a real transformer.

We again begin by expressing v_m as a function of v_1. However, this time the function is a little more complex because of the series inductor:

$$v_m = \frac{[R_M||(1/sC_M)]}{sL + [R_M||(1/sC_M)]} N_M v_1. \tag{4.48}$$

Next we express v_1 as a function of v_s:

$$v_1 = \frac{\dfrac{1}{N_M^2}[sL + R_M||(1/sC_M)]}{R_S + \dfrac{1}{N_M^2}[sL + R_M||(1/sC_M)]} v_s. \tag{4.49}$$

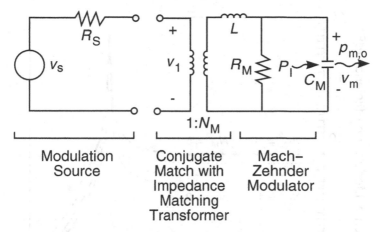

Figure 4.25 Schematic of a conjugate match between a resistive source and the
RC impedance model for a Mach–Zehnder modulator.

Expanding out the parallel impedances in (4.48) and (4.49), substituting (4.49)
into (4.48) and then substituting the result into (4.39) yields an expression for the
modulated optical power from the Mach–Zehnder as a function of the modulating
voltage:

$$
p_{m,o} = \left(\frac{T_{FF} P_1 \pi}{2 V_\pi} \right) \left(\frac{N_M^2 R_S}{N_M^2 R_S + R_M} \right)
$$

$$
\times \frac{1}{\left(s^2 \dfrac{R_M L C_M}{N_M^2 R_S + R_M} + s \dfrac{L + N_M^2 R_S R_M C_M}{N_M^2 R_S + R_M} + 1 \right)} v_s. \qquad (4.50)
$$

To obtain the incremental modulation efficiency from (4.50), we assume the transformer turns ratio is chosen such that $N_M^2 R_S = R_M$. Next we square (4.50) and
divide by the source available power and group the dc terms according to the
Mach–Zehnder slope efficiency, (4.44); the result of all these steps is

$$
\left. \frac{p_{m,o}^2}{p_{s,a}} \right|_{\substack{\text{conjugate} \\ \text{match}}} = \frac{s_{mz}^2 N_M^2}{R_S} \frac{1}{\left(s^2 \dfrac{L C_M}{2} + s \dfrac{L + R_M^2 C_M}{2 R_M} + 1 \right)^2}. \qquad (4.51)
$$

Like the incremental modulation efficiencies with conjugate match for the photodiode and diode laser, (4.51) has a bandpass type of frequency response. To
see the effects of various impedance matching ratios, we have plotted (4.51) in
Fig. 4.26 using $L = 2$ nH and a representative range of N_M. Because we constrained

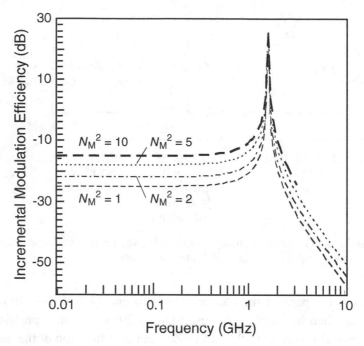

Figure 4.26 Modeled incremental modulation efficiency vs. frequency for a conjugate match to a Mach–Zehnder modulator with the transformer turns ratio as a parameter.

the inductor loss to be that value needed for matching, we obtained a relatively high Q resonance, which in turn accounts for the sharp peaking in the frequency response.

4.3.3.3 External modulation link with passive impedance matching

Now that all the hard work is done to develop expressions for the incremental modulation efficiency of a Mach–Zehnder – and previously for a photodiode – it is time to harvest some of it to develop expressions for the frequency response of an impedance-matched external modulation link. We again choose to form link expressions for the unmatched and conjugately matched cases. The schematic diagrams of the four links that result from these combinations are given in Fig. 4.27.

Using once more the property that the link frequency response is simply the product of the frequency responses of the modulation and photodetection devices' incremental modulation and detection efficiencies, we can write the desired results as the pairwise combinations of either (4.43) or (4.51) together with either (4.12)

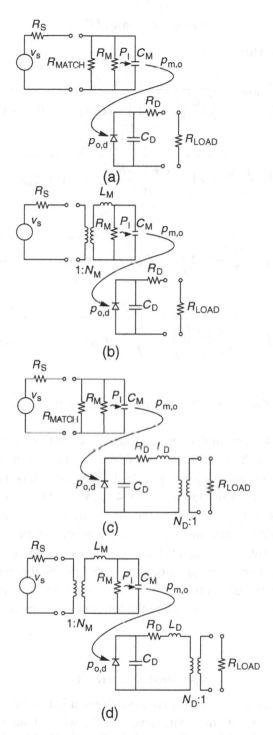

Figure 4.27 Schematic diagrams for the externally modulated links with: (a) no matching to either the Mach–Zehnder or the photodiode; (b) conjugate matching to the Mach–Zehnder but no matching to the photodiode; (c) no matching to the Mach–Zehnder but conjugate matching to the photodiode; and (d) conjugate matching to both the Mach–Zehnder and the photodiode.

where $N_D = 1$ or (4.16):

$$g_{i,ummz,umd} = \frac{s_{mz}^2 r_d^2}{\left(\dfrac{sC_M R_S}{2} + 1\right)^2 (sC_D(R_D + R_S) + 1)^2}, \qquad (4.52)$$

$$g_{i,mmz,umd} = \frac{s_{mz}^2 r_d^2 N_M^2}{\left(s^2 \dfrac{C_M L_M}{2} + s\left(\dfrac{L_M + C_M R_M^2}{2R_M}\right) + 1\right)^2 (sC_D(R_D + R_S) + 1)^2}, \qquad (4.53)$$

$$g_{i,ummz,md} = \frac{s_{mz}^2 r_d^2 N_D^2}{\left(\dfrac{sC_M R_S}{2} + 1\right)^2 \left(s^2 C_D L_D + sC_D\left(R_D + N_D^2 R_S\right) + 1\right)^2}, \qquad (4.54)$$

$$g_{i,mmz,md} =$$
$$\frac{s_{mz}^2 r_d^2 N_M^2 N_D^2}{\left(s^2 \dfrac{C_M L_M}{2} + s\left(\dfrac{L_M + C_M R_M^2}{2R_M}\right) + 1\right)^2 \left(s^2 C_D L_D + sC_D\left(R_D + N_D^2 R_S\right) + 1\right)^2}, \qquad (4.55)$$

where we have added subscripts to distinguish the modulator and photodiode inductances.

Experimental links that implement the four cases listed above have been assembled and measured by Prince (1998). Plots of the calculated and measured intrinsic gain vs. frequency are presented in Fig. 4.28. In comparing this figure with the corresponding one for direct modulation, Fig. 4.21, we note the key similarity: that the intrinsic gain is simply the sum of the matching gains for the modulation and photodetection devices. This similarity persists despite the differences in modulation device, center frequency, bandwidth, etc. Of course there are differences as well; prime among them is the broadband gain that is possible with external modulation – which can be substantial, 12 dB in this case – and which is not feasible in direct modulation links at present due to the lower slope efficiency of current direct modulation devices.

4.4 Bode–Fano limit

Following the discussions in the previous sections, it is natural to ask: what is the best impedance match that one can achieve over the widest bandwidth? The classic answer to this question is given by the early work of Bode (1945), which was generalized by Fano (1950). Consequently the limit that resulted from these two efforts is generally, although not universally, referred to as the Bode–Fano limit.

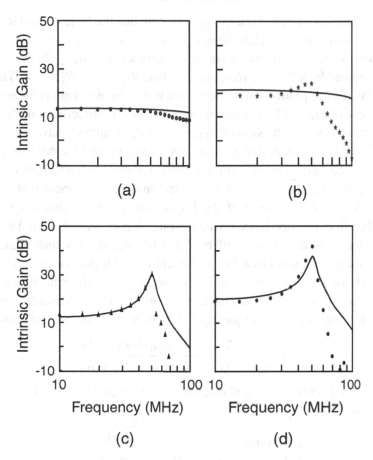

Figure 4.28 Plots of the intrinsic gain vs. frequency for the four circuits presented in Fig. 4.27: (a) no matching to either the Mach–Zehnder or the photodiode; (b) conjugate matching to the Mach–Zehnder but no matching to the photodiode; (c) no matching to the Mach–Zehnder but conjugate matching to the photodiode; and (d) conjugate matching to both the Mach–Zehnder and the photodiode. (Prince, 1998, reprinted with permission from the author.)

This limit only applies to lossless impedance matching. To appreciate the basis for this restriction, we first consider lossy impedance matching.

4.4.1 Lossy impedance matching

We begin our discussion of the tradeoffs that impedance matching introduces between gain and bandwidth by considering matching via *lossy* circuits. We choose to analyze the resistive match to a diode laser as an example of this set of matching circuits.

For the initial part of this discussion we assume that the laser is ideal in the sense that it has no capacitance. Therefore, the circuit with which we deal initially is shown in Fig. 4.15, but without the laser capacitance. For $R_L \leq R_S$ an impedance match is obtained by selecting R_{MATCH} such that $R_{MATCH} + R_L = R_S$. There are a couple of interesting points to note here. One is that the link gain is independent of R_L, as we saw in Fig. 4.17(a). This observation follows directly from the fact that the laser current is set by the sum of R_{MATCH} and R_L, which we have agreed to hold constant. The other point is that we have achieved a perfect match over an infinite bandwidth – a consequence of assuming there was no laser capacitance.

Now suppose that we do not have a perfect match, i.e. suppose that $R_{MATCH} + R_L \neq R_S$. We need a measure of the impedance mismatch that can be used to calculate the effect of impedance mismatch on link gain. The reflection coefficient, Γ, which quantifies the fraction of reflected RF signal, is a such a measure of impedance mismatch, and has a further benefit in that it plays a central role in the Bode–Fano limit discussion in the next section. In general the reflection coefficient has both a magnitude and phase. However, in this section we consider only purely resistive components and consequently Γ is real. In this case we can define Γ as

$$\Gamma \equiv \frac{R_{IN} - R_S}{R_{IN} + R_S} = \frac{R_{MATCH} + R_L - R_S}{R_{MATCH} + R_L + R_S}, \tag{4.56}$$

where R_{IN} is the purely real impedance seen by the source resistor. To get a feel for Γ, we note the following special cases:

$$R_{MATCH} + R_L = R_S \leftrightarrow \Gamma = 0, \tag{4.57a}$$

$$R_{MATCH} + R_L \gg R_S \leftrightarrow \Gamma \cong +1, \tag{4.57b}$$

$$R_{MATCH} + R_L \ll R_S \leftrightarrow \Gamma \cong -1. \tag{4.57c}$$

Since we want to explore the effects of impedance mismatch on link gain, we need to start with (2.10), which expresses the modulated optical power from a diode laser with an arbitrary value for the series matching resistor; this equation is repeated below for convenience:

$$p_{\ell,o}^2 = \frac{s_\ell^2 v_s^2}{(R_{MATCH} + R_L + R_S)^2}. \tag{4.58}$$

To obtain an expression for the intrinsic gain from (4.58) we need to multiply by the incremental detection efficiency of the photodiode, (2.38), and divide the result by the source available power, (2.7); the result is

$$g_{i,\text{unmatched}} = s_\ell^2 r_d^2 \frac{4 R_s^2}{(R_{MATCH} + R_L + R_S)^2}, \tag{4.59}$$

where we have made the assumptions that the source and load resistances are equal, and that there is negligible optical loss between laser and detector.

We are interested in exploring the effects of impedance mismatch on link gain. Thus we normalize (4.59) by the gain with an impedance match, viz.:

$$g_{i,\text{matched}} = s_\ell^2 r_d^2. \tag{4.60}$$

Dividing (4.59) by (4.60) yields an expression for the normalized link gain:

$$g_{i,\text{normalized}} = \frac{g_{i,\text{unmatched}}}{g_{i,\text{matched}}} = \frac{4R_s^2}{(R_{\text{MATCH}} + R_L + R_S)^2}. \tag{4.61}$$

Dividing the numerator and denominator of (4.56) by $R_{\text{MATCH}} + R_L$ and (4.61) by $(R_{\text{MATCH}} + R_L)^2$ produces expressions that both depend on the ratio $R_S/(R_{\text{MATCH}} + R_L)$:

$$\Gamma = \frac{1 - \dfrac{R_S}{R_{\text{MATCH}} + R_L}}{1 + \dfrac{R_S}{R_{\text{MATCH}} + R_L}}, \tag{4.62}$$

$$g_{i,\text{normalized}} = \frac{4\left(\dfrac{R_S}{R_{\text{MATCH}} + R_L}\right)^2}{\left(1 + \dfrac{R_n}{R_{\text{MATCH}} + R_L}\right)^2}. \tag{4.63}$$

Figure 4.29 is a plot of (4.63) vs. (4.62) with $R_S/(R_{\text{MATCH}} + R_L)$ as the parameter. As expected, when there is a perfect match, i.e. when $R_{\text{MATCH}} + R_L = R_S$, the reflection coefficient equals 0, (4.62); and the intrinsic gain equals the matched intrinsic gain, (4.63).

When the matching resistor is larger than the match value, the intrinsic gain decreases from the matched case. Conversely, when the matching resistor is less than the match value, the intrinsic gain actually increases over the match value. From (4.63) we can see that the gain increase available from this method is at most a factor of 4, or 6 dB. Thus if maximum intrinsic gain was the only parameter of interest, then setting the match resistor to zero and designing a laser with zero resistance would be optimum.

However, as Fig. 4.29 also indicates, selecting a match resistor value such that $R_{\text{MATCH}} + R_L \neq R_S$ – whether larger or smaller than the match value – results in a fraction of the incident power from the source being reflected by the laser load resistance back to the source. This can be a problem for some sources; in other cases it increases the distortion because of the reflected waves set up by the

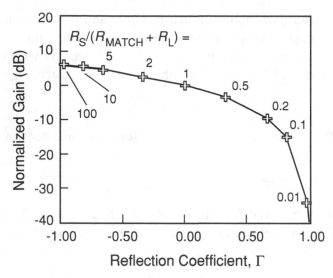

Figure 4.29 Plot of the normalized gain vs. the reflection coefficient with the degree of match as a parameter for a link with a series resistance (lossy) impedance match.

impedance mismatch. Despite these drawbacks, omitting the matching resistor, i.e. setting $R_{MATCH} = 0$, is often done in practice.

In summary, the tradeoffs among gain, bandwidth and reflection coefficient for a particular lossy match implementation can be determined by calculating for *each* case equations such as (4.62) and (4.63), which we derived for the series resistive match case. At present there is no general expression that relates these parameters, independent of the lossy match topology. The reason we lack such a general relationship is that with a lossy matching circuit there are too many places for the power to go – into the load, reflected to the source, or absorbed by the matching circuit.

4.4.2 Lossless impedance matching

A lossless matching circuit narrows the possible places for the power to go by eliminating the matching circuit loss. Therefore with a lossless impedance match, the power can only be absorbed by the load or be reflected back to the source. This seemingly trivial simplification is enough to permit the derivation of two general relationships that hold for all lossless matching circuits, regardless of topology. One relates the load impedance to the bandwidth over which a given degree of match, as measured by the magnitude of the reflection coefficient, is achieved; the other general expression relates the gain in terms of the reflection coefficient.

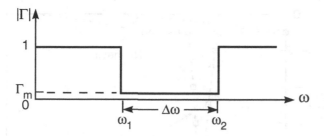

Figure 4.30 Plot of the reflection coefficient vs. frequency that gives the widest bandwidth for a given degree of lossless impedance match, as measured by the reflection coefficient. (Pozar, 1993, Fig. 6.23a. © 1993 Pearson Education, Inc., reprinted with permission.)

Since in the following we need to expand the set of matching components to include reactive elements, we also need to expand the reflection coefficient to be complex. Of the several equivalent ways to represent a complex number, the most common for reflection coefficient is to use magnitude and phase: $\Gamma(\omega) = \rho(\omega)\,e^{j\theta(\omega)}$.

We begin with the relationship between the degree of match and the bandwidth. Although this relationship covers all lossless matching circuits, its detailed form does depend on the load impedance circuit. Consider first a parallel RC load impedance. Bode (1945) first showed that the reflection coefficient of the load as viewed through a lossless matching circuit must satisfy the following inequality:

$$\int_0^\infty \ln\frac{1}{\rho(\omega)}\,d\omega \le \frac{\pi}{RC}. \tag{4.64}$$

A rigorous derivation of this superficially simple relationship is lengthy and involved; therefore we do not repeat it here but refer the interested reader to the original works of Bode (1945) and Fano (1950) or to the somewhat more readable derivation of Carlin (1954).

Fano also demonstrated that the most effective shape for the reflection coefficient is that shown in Fig. 4.30. This reflection coefficient reflects all the power outside of a band $\Delta\omega = \omega_2 - \omega_1$ while passing most, but not all, of the power over the bandwidth $\Delta\omega$. Using this form of reflection coefficient in (4.64), the frequency spans for which $\rho(\omega) = 1$ do not contribute to the integral since $\ln(1) = 0$. Thus the only contribution to the integral in (4.64) is over the band $\Delta\omega$. Consequently for the reflection function shown in Fig. 4.30, equation (4.64) evaluates to

$$\Delta\omega \ln\frac{1}{\rho_m} \le \frac{\pi}{RC}. \tag{4.65}$$

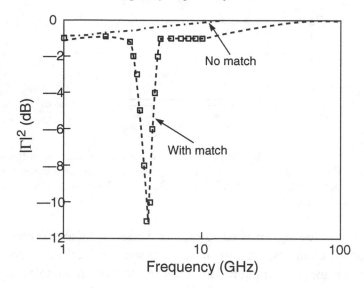

Figure 4.31 Plot of $|\Gamma|^2$ vs. frequency for a diode laser with no impedance matching (dash-dot line) and with impedance matching (squares are experimentally measured values of Onnegren and Alping *et al.*, 1995).

This form of the Bode–Fano limit clearly shows that there is a tradeoff between the bandwidth and the reflection coefficient: the better the impedance match – i.e. the smaller ρ_m – the narrower the bandwidth over which this degree of impedance match can be obtained. It is interesting to note that in the limit of a perfect match, i.e. $\rho_m = 0$, equation (4.65) shows this is possible only over an infinitesimally small bandwidth. The practical implication of (4.65) for designing a lossless matching circuit is that one should only strive for the degree of match required, because achieving a better match only further limits the bandwidth.

We can take the analysis one step further and calculate the optimum functional form of the reflection coefficient for maximum bandwidth given a particular value of reflection coefficient. This is easily done by solving (4.65) when the equality holds for ρ_m:

$$\rho_m = e^{-\pi/\Delta\omega RC}. \tag{4.66}$$

Unfortunately it turns out that a network satisfying (4.66) cannot be synthesized with a finite number of linear passive elements. However, this expression can serve as a goal for the matching circuit designer to approximate as closely as possible.

To get a feel for what can be achieved in practice, we now evaluate the integral in (4.64) for the example of a resonant match to the diode laser presented in Section 4.3.2.2. Figure 4.31 presents plots of $|\Gamma|^2 = \rho^2$ vs. frequency for two

Table 4.1 *Calculated gain–BW values and comparison to Bode–Fano limit*

Case	Value of integral in (4.64)
Without impedance match	3.5×10^9 s^{-1}
With impedance match	1.3×10^{10} s^{-1}
Bode–Fano limit	6.8×10^{10} s^{-1}

cases. One curve is the calculated reflection coefficient based on the measured laser impedance alone, i.e. without any matching. The other curve in Fig. 4.31 is the measured reflection coefficient of the same laser conjugately matched to the resistive source.

Using numerical integration, Prince (1998) has evaluated (4.64) for these two cases. The results are listed in Table 4.1, along with the Bode–Fano limit for this laser impedance with perfect impedance matching. As we can see from these results, the case without any impedance matching between source and laser falls about a factor of 20 short of the Bode–Fano limit value with perfect matching. This quantifies our intuition that one pays a significant penalty for connecting the laser impedance directly to the source without any impedance matching. Using the data from a representative experimental impedance matching circuit, it is possible to improve on the unmatched case and come within a factor of 5 of the Bode–Fano limit.

As mentioned above, although the Bode–Fano limit applies to all lossless matching circuits, the particular limit in (4.64) depends on the load circuit, as was first pointed out by Fano (1950). Since we can often approximate the impedances of most electro-optic devices via either a series or parallel connection of a resistor and either a capacitor or inductor, there are really only four limits needed. These four load impedances, along with their corresponding Bode–Fano limits, are listed in Fig. 4.32 (Pozar, 1993).

The other general relationship that holds for a lossless network is between the gain and the reflection coefficient. Carlin (1954), among others, has shown this relationship to be

$$g^2 = (1 - |\Gamma_1|^2) = 1 - \rho_1^2. \tag{4.67}$$

Like the Bode–Fano limit, this relationship follows directly from the lossless condition, since for such networks any power that is not reflected is transmitted – which is what (4.67) indicates.

We now have the tools needed to investigate the Bode–Fano limits on the gain and bandwidth of passively matched directly and externally modulated links. We demonstrate these tools by applying them to the direct modulation case (Gulick

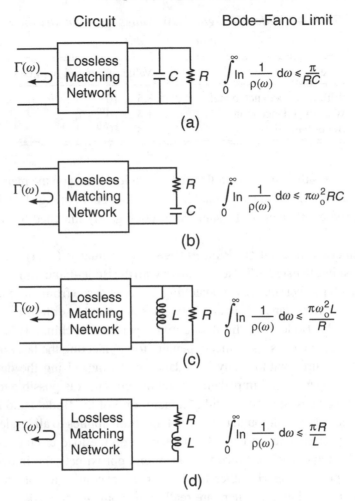

Figure 4.32 Bode–Fano limits for four common load impedances: (a) parallel RC, (b) series RC, (c) parallel RL, and (d) series RL. (Pozar, 1993, Fig. 6.22. © 1993 Pearson Education, Inc., reprinted with permission.)

et al., 1986). The link they analyzed is shown in Fig. 4.33; it is basically a combination of the parallel RC laser model, the series RC photodiode model with the addition of two lossless matching circuits, M_1 and M_2 for matching the source to the laser, and the photodetector to the load, respectively. Gulick *et al.* have shown that the gain for this link can be expressed in terms of the circuit parameters and the input and output reflection coefficients (by using (4.67) twice) as

$$g_i = \frac{s_\ell^2 r_d^2}{4\omega^2 R_L R_D C_D^2}\left(1 - \rho_1^2\right)\left(1 - \rho_2^2\right). \tag{4.68}$$

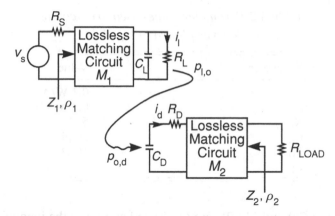

Figure 4.33 Direct modulation link with lossless impedance matching circuits for the laser and photodetector. (Cox and Ackerman, 1999, Fig. 4. © 1999 International Union of Radio Science, reprinted with permission.)

The first term in (4.68) is the gain on resonance, whereas the latter two terms use (4.67) twice – once at the link input and once at the link output – to modify the on-resonance gain for off-resonance frequencies.

We already have an expression for the laser matching circuit reflection coefficient, ρ_1, in (4.66). The photodiode reflection coefficient can be determined in an analogous manner by solving (4.64) using the series form of the Bode–Fano limit from Fig. 4.32(b):

$$\rho_2 = e^{-\pi R_D C_D \omega_0^2 / \Delta\omega}, \tag{4.69}$$

where $\omega_0 = \sqrt{\omega_1 \omega_2}$, i.e. the resonant frequency is the geometric mean of the two band edge frequencies.

An important special case of (4.68) occurs when the laser reflection coefficient is much less than the photodiode reflection coefficient and consequently can be ignored. This condition corresponds to the situation where the exponent in (4.66) is larger, in absolute value, than the exponent in (4.69). Writing out this inequality between these two exponents, canceling the common terms and moving all the electrical component values to the same side of the inequality yields

$$\frac{1}{(R_L C_L)(R_D C_D)} > \omega_0^2. \tag{4.70}$$

The parentheses have been added in (4.70) to emphasize that the laser reflection coefficient can be ignored when the center of the matching passband is less than the geometric mean of the laser and photodiode 3-dB frequencies.

Table 4.2 *Parameter values used to calculate*
the curves plotted in Fig. 4.34

Parameter	Value
s_l	0.3 W/A
r_d	0.7 A/W
R_L	5 Ω
C_L	20 pF
R_D	10 Ω
C_D	0.5 pF

Substituting (4.66) and (4.69) into (4.68) and assuming the inequality in (4.70) is satisfied, the approximate expression for link gain becomes

$$g_i \cong \frac{s_l^2 r_d^2}{4\omega_0^2 R_L R_D C_D^2}\left(1 - e^{-\pi R_D C_D \omega_0/\delta}\right),$$
(4.71)

where $\delta = \frac{\omega_2 - \omega_1}{\omega_0}$ is the fractional bandwidth. Equation (4.71) expresses the relationship between direct modulation link gain and the fractional bandwidth, assuming ideal lossless matching.

To get a feel for the link gains and fractional bandwidths that are theoretically possible, we assume the device parameter values listed in Table 4.2, which are near the top of the range reported for present diode lasers and high speed photodiodes at 1.3 μm. From the values listed in Table 4.2, the geometric mean of the laser and photodiode 3-dB frequencies is 7.1 GHz; consequently according to (4.70) the maximum center frequency needs to be kept below this value.

Figure 4.34 is a plot of (4.71) vs. fractional bandwidth with matching passband center frequency as a parameter.[10] These curves represent the Bode–Fano limits on the intrinsic gain and bandwidth of a direct modulation link with lossless passive matching. As expected from the tradeoff between gain and bandwidth imposed via this limit, the link gain decreases as the fractional bandwidth and/or passband center frequency increase. And conversely as the bandwidth and/or center frequency decrease the intrinsic gain increases.

However, what is typically not expected on first examination is that for sufficiently low values of these same two parameters, a direct modulation link is capable of positive intrinsic gain. The question that is often asked is: in this collection of seemingly passive components, where is the power gain coming from? After all, a

[10] It is important to note that although equations such as (4.71) are expressed in terms of ω, i.e. radian frequencies, in plotting Fig. 4.34 we have converted all these frequencies, whose units are radians per second, into cycles per second, whose units are Hz.

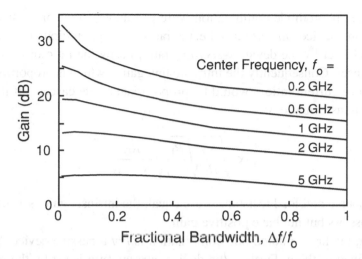

Figure 4.34 Bode–Fano limits on the intrinsic link gain of a directly modulated link with ideal lossless, passive impedance matching vs. fractional bandwidth of the matching circuit with the center frequency of the matching passband as a parameter.

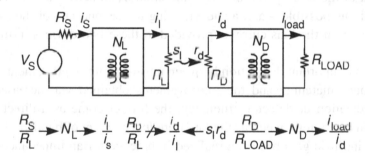

Figure 4.35 Block diagram of a directly modulated link with lossless transformer matching to the laser and photodetector.

transformer – which we have seen is a reasonable way to implement lossless passive impedance matching – can have voltage *or* current gain, but not power gain.

To assist in providing a rationale for the positive gain, consider the link shown in Fig. 4.35. The laser has a resistance less than the source resistance, while the photodetector has a resistance greater than the load resistance. The impedance matching transformers take care of matching the source and load to the laser and photodiode respectively. These source-to-laser and photodiode-to-load turns ratios, which are determined by the ratios of the source-to-laser and photodiode-to-load resistances, set the ratios of the laser to source and load to photodiode currents. To put the previous point succinctly: the source-to-laser and photodiode-to-load *current* ratios are set by the corresponding source and load *resistance* ratios.

If the laser–photodiode combination were just another pair of passive components, then the device *current* ratio, i.e. the ratio of the photodiode to laser current would also be set by the device *resistance* ratio, i.e. by the ratio of these two device resistances. Consequently the intrinsic link gain – which is proportional to the load-to-source current ratio – would be proportional to the cascade of these three impedance transformations:

$$g_i \propto \sqrt{\frac{R_S}{R_L}} \sqrt{\frac{R_L}{R_D}} \sqrt{\frac{R_D}{R_{LOAD}}}. \tag{4.72}$$

When the source and load resistances are equal, the intrinsic gain is 1; i.e. the link could be lossless but not have positive gain.

However, the laser–photodiode combination is *not* a passive device: both have dc bias applied to them. Further, the device *current* ratio is set by the product of the *slope efficiency* and *responsivity* of these two devices – not the device *resistance* ratio. Consequently the device current ratio can be larger than the device resistance ratio, which is the middle ratio in (4.72). Therefore it is more appropriate to think of the laser–photodiode combination as an active device. The power to make all this possible – and avoid violating the second law of thermodynamics – comes from the bias currents provided to these two devices, primarily the laser.

There is an important distinction between intrinsic gain improvements obtained via impedance matching and the improvements obtained via increased optical power, modulation or detector efficiency: the former come as a direct result of a tradeoff in bandwidth, whereas the latter are *independent* of bandwidth.

Positive intrinsic gain has been realized in practice for an impedance-matched direct modulation link, despite the fact that realizable impedance matching components have loss. Although Cox (1986) did an early demonstration of this effect, Ackerman *et al.* (1990) were the first to demonstrate a direct modulation link with gain at an interesting frequency. Ackerman *et al.* built a link that had a peak gain of 3.7 dB over a bandwidth of 90 MHz, centered at 900 MHz.

It is instructive to compare the results of practical implementations of impedance matching to their corresponding Bode–Fano limits. To do this we have selected three examples from the literature (Ackerman *et al.*, 1990; Goldsmith and Kanack, 1993; Onnegren and Alping, 1995) and listed the values needed for the calculation in Table 4.3. In Fig. 4.36 we plot the calculated and experimentally measured values for these three cases.

From these data points we see that the narrow bandwidth experimental results (around 10% fractional bandwidth) come quite close to the corresponding

Table 4.3 *Parameter values for three experimental links used to calculate their gain–BW products, as plotted in Fig. 4.36*

Parameter	Ackerman *et al.*	Goldsmith and Kanack	Onnegren and Alping
s_ℓ (W/A)	0.11	0.1	0.015
r_d (A/W)	1.0	0.85	1
R_L (Ω)	5.9	4.25	3.9
C_L (pF)	11.3	13	58
R_D (Ω)	15.8	5	12.5
C_D (pF)	0.35	0.37	0.46
ω_0 (radians/s)	5.65×10^9	1.88×10^{10}	2.58×10^{10}
$\Delta\omega$ (radians/s)	5.65×10^8	1.26×10^{10}	2.58×10^9
δ	0.1	0.67	0.1

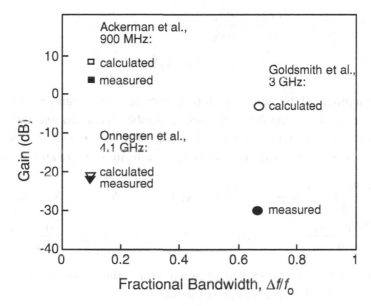

Figure 4.36 Calculated and measured Bode–Fano limits for the three impedance matched links listed in Table 4.3.

Bode–Fano limiting values, essentially independent of center frequency. However, as one attempts to increase the bandwidth of the match, we see the experimental results fall far short of what is predicted theoretically. This discrepancy is not the result of sloppy experimental work. Rather it is a testament to the difficulty of realizing – with present components – the theoretical potential that the Bode–Fano limit permits in wide bandwidth applications.

Appendix 4.1 Small-signal modulation rate equation model for diode lasers

In Appendix 2.1, we introduced the rate equations that describe a semiconductor laser as two coupled reservoirs of electrons and photons. In that appendix, the steady state or dc solutions were found. Here we are concerned with the dynamics and derive the small-signal modulation response. The approach presented here is based on the development of Coldren and Corzine (1995) with the discussion of the dynamic solution to the rate equations provided by Lee (1998).

Because the laser is biased well above threshold in virtually all analog applications, we can simplify the rate equations for this analysis by making the following two observations. In the photon rate equation, stimulated emission dominates over spontaneous emission giving $r_{ST} \gg \beta_{SP}\tau_{SP}$; and in the carrier rate equation, carrier recombination is dominated by stimulated emission giving $v_g g n_P \gg n_U/\tau_U$. Therefore we can neglect these terms in (A2.1.14a and b); the results are

$$\frac{dn_U}{dt} = \frac{\eta_i i_L}{q V_E} - v_g g(n_U, N_P) n_P, \tag{A4.1a}$$

$$\frac{dn_P}{dt} = \Gamma v_g g(n_U, N_P) n_P - \frac{n_P}{\tau_P}. \tag{A4.1b}$$

The development of the small-signal frequency response begins by following our usual practice of expressing the laser current, number of carriers and cavity photons each as the sum of a dc term and a small-signal term, viz.: $i_L = I_L + i_\ell$, $n_U = N_U + n_u$ and $n_P = N_P + n_p$. Substituting these expressions into (A4.1a and b), we obtain

$$\frac{d(N_U + n_u)}{dt} = \frac{\eta_i(I_L + i_\ell)}{q V_E} - v_g g(N_U + n_u, N_P + n_p)(N_P + n_p); \tag{A4.2a}$$

$$\frac{d(N_P + n_p)}{dt} = \Gamma v_g g(N_U + n_u, N_P + n_p)(N_P + n_p) - \frac{N_P + n_p}{\tau_p}. \tag{A4.2b}$$

The gain function can be expanded in a Taylor series of the first order since n_u and n_p are very small. Thus, we can substitute

$$g(N_U + n_u, N_P + n_p) \approx g(N_U, N_P) + a_N n_u - a_P n_p, \tag{A4.3}$$

where $a_N = \partial g/\partial n_U$ and $a_P = -\partial g/\partial n_P$. In the steady state, the rate of change of the sum of the dc terms in each of the rate equations must be zero. Using this fact and substituting (A4.3) into (A4.2a and b) we obtain

$$\frac{dN_U}{dt} = 0 = \frac{\eta_i I_L}{q V_E} - v_g g(N_U, N_P) N_P, \tag{A4.4a}$$

$$\frac{dN_P}{dt} = 0 = \Gamma v_g g(N_U, N_P) N_P - \frac{N_P}{\tau_P}. \tag{A4.4b}$$

Removing the dc terms – (A4.4a and b) – from (A4.2a and b), respectively, leaves us with expressions for the small-signal rate equations:

$$\frac{dn_u}{dt} = -v_g a_N N_P n_u - (v_g g(N_U, N_P) - v_g a_P N_P) n_p$$

$$+ \frac{\eta_i i_\ell}{q V_E} - v_g a_N n_u n_p + v_g a_P n_p^2, \tag{A4.5a}$$

$$\frac{dn_p}{dt} = \Gamma v_g a_N N_P n_u + \left(\Gamma v_g g(N_U, N_P) - \frac{1}{\tau_p} - \Gamma v_g a_P N_P \right) n_p$$

$$+ \Gamma v_g a_N n_u n_p - \Gamma v_g a_P n_p^2. \tag{A4.5b}$$

We can make three additional simplifications:[11] (1) recall that the gain is clamped at the threshold value, $\Gamma g_{TH} \approx \text{loss} = \frac{1}{v_g \tau_p}$, i.e. the addition or subtraction of one photon does not change the threshold condition; (2) neglect the gain compression terms by setting $a_P = 0$, which means that our results will underestimate the damping; and (3) neglect the small non-linear $n_u n_p$ and n_p^2 in linear response calculations (we will need to reinstate these terms when we derive the non-linear distortion properties of the lasers (Appendix 6.1)). This leaves us with a coupled pair of *linear* differential equations in n_u and n_p:

$$\frac{dn_u}{dt} = -v_g a_N N_P n_u - \frac{1}{\Gamma \tau_p} n_p + \frac{\eta_i i_\ell}{q V_E}, \tag{A4.6a}$$

$$\frac{dn_p}{dt} = \Gamma v_g a_N N_P n_u. \tag{A4.6b}$$

At this point we have arrived at the general form of the small-signal rate equations that we will use to derive the dynamic modulation responses of the laser. Before going through the mathematics, let us take a more careful look at the physics in equations (A4.6a and b). Ignoring the driving terms, (A4.6a) really only expresses the rate of change of the carrier density due to changes in the stimulated emission. The first term in (A4.6a) $v_g a_N N_P n_u$ is the homogeneous part of the equation and represents a decay in the carrier density due to increased stimulated emission from the differential increase in the gain, i.e. this is the damping term. The second term in (A4.6a), $n_p / \Gamma \tau_p$, is the coupled part of the equation and represents a linear decay of the carrier density due to increased stimulated emission from an increase in the photon density. Because the carrier density decreases when the photon density increases – and vice versa in equation (A4.6b) – we recognize that the coupled term

---

[11] Readers concerned with the simplifications we have made, may refer to Coldren and Corzine (1995) for a comprehensive treatment of the rate equations. We are justified in doing away with so many terms because the results agree fairly well with experiment. Thus these simplified rate equations allow us to build our intuition for laser dynamics.

in equation (A4.6b) is from the decay term in (A4.6a). Consequently these coupled terms can yield oscillations if the damping is not too high.

Another important point is that since the damping term in (A4.6a) is the coupled term in (A4.6b), the damping increases as the bandwidth increases. Consequently it is not possible to achieve arbitrarily high modulation bandwidths because once the damping exceeds a certain value, it limits further bandwidth increases. In practice it is often the case that other phenomena such as device heating and parasitic impedances are more important in limiting the bandwidth.

Now that we know what to expect, let us solve the rate equations to derive the modulation response quantitatively. A common way to solve these equations is to take the Fourier transforms of (A4.6a and b), which converts the differential equations into algebraic equations:

$$j\omega n_u = -v_g a_N N_P n_u - \left(\frac{1}{\Gamma \tau_p}\right) n_p + \frac{\eta_i i_\ell}{q V_E}, \tag{A4.7a}$$

$$j\omega n_p = \Gamma v_g a_N N_P n_u. \tag{A4.7b}$$

We can resort to any of the numerous techniques available to solve this system of linear algebraic equations. A compact method of solution is to write (A4.7a and b) using matrix notation and solve by matrix inversion, viz.:

$$\begin{bmatrix} v_g a_N N_P + j\omega & \dfrac{1}{\Gamma \tau_p} \\ -\Gamma v_g a_N N_P & j\omega \end{bmatrix} \begin{bmatrix} n_u \\ n_p \end{bmatrix} = \begin{bmatrix} \dfrac{\eta_i i_\ell}{q V_E} \\ 0 \end{bmatrix}. \tag{A4.8}$$

To solve for n_u and n_p we can pre-multiply both sides of (A4.8) with the inverse of the matrix on the left. Recall the formula for inverting a 2×2 matrix is given by

$$\begin{bmatrix} A & B \\ C & D \end{bmatrix}^{-1} = \frac{1}{AD - BC} \begin{bmatrix} D & -B \\ -C & A \end{bmatrix}. \tag{A4.9}$$

Applying (A4.9) to (A4.8) gives us the solution

$$\begin{bmatrix} n_u \\ n_p \end{bmatrix} = \frac{H(j\omega)}{\omega_r^2} \begin{bmatrix} j\omega & -\dfrac{1}{\Gamma \tau_p} \\ \Gamma v_g a_N N_P & v_g a_N N_P + j\omega \end{bmatrix} \begin{bmatrix} \dfrac{\eta_i i_\ell}{q V_E} \\ 0 \end{bmatrix} \tag{A4.10}$$

where the pre-factor is the determinant of the matrix in (A4.8) that is given by

$$\frac{H(j\omega)}{\omega_r^2} = \frac{1}{-\omega^2 + j\omega(v_g a_N N_P) + \dfrac{v_g a_N N_P}{\tau_p}}. \tag{A4.11}$$

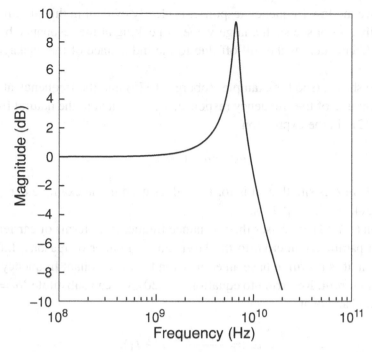

Figure A4.1 Plot of (A4.14) – the small-signal frequency response of a diode laser – using the values listed immediately following (A4.14). (Lee, 1998, reprinted with permission of the author.)

It is common to rewrite (A4.11) in terms of a natural or resonant frequency, ω_r, and a damping factor, ζ, which are the commonly used variables for describing a second-order frequency response. In this case if we define these variables as follows:

$$\omega_r = \sqrt{\frac{v_g a_N N_P}{\tau_p}}, \tag{A4.12}$$

$$\zeta = \frac{\gamma}{2\omega_r} = \sqrt{\tau_p v_g a_N N_P}, \tag{A4.13}$$

then we can rewrite (A4.11) in the following standard form:

$$H(j\omega) = \frac{1}{\left[1 - \left(\dfrac{\omega}{\omega_r}\right)^2\right] + j2\zeta\dfrac{\omega}{\omega_r}}. \tag{A4.14}$$

For typical values of v_g, a_N, N_P, and τ_p (Coldren and Corzine, 1995; Table 5.1) we can evaluate (A4.12) and (A4.13) to obtain representative values of ω_r and ζ, respectively: $\omega_r = 41.1 \times 10^9$ radians/s or $f_r = 6.54\,\text{GHz}$; $\zeta = 0.057$. Figure A4.1 is a plot of (A4.14) using these values. The relaxation resonance peak, which is almost

10 dB above the low frequency response, is clearly evident in this plot. In practice we typically do not see such a large value of peaking at the resonance because of parasitic effects such as the roll-off due to the inductance of the packaging bond wire.

It can be shown (see for example Roberge, 1975) that the frequency at the peak in the magnitude of the frequency response, ω_p, is related to the natural frequency, ω_r of (A4.12) via the expression

$$\omega_p^2 = \omega_r^2(1 - 2\zeta)^2. \tag{A4.15}$$

In cases where ζ is small, as it is for the values used in the example here, then we see from (A4.15) that $\omega_p \cong \omega_r$.

Equation (A4.12) expresses the resonance frequency in terms of carrier density, which is a parameter internal to the laser. For a number of reasons, link design among them, it is useful to have an expression for the resonant frequency in terms of the laser current. Referring to equation (A2.20), we can substitute $N_P = (I - I_T)$ $\tau_p \eta_i / q V_p$ and arrive at

$$\omega_r^2 = \frac{v_g a_N \Gamma \eta_i}{q V_P}(I - I_T), \tag{A4.16}$$

and for the damping

$$\zeta = \frac{\gamma}{2\omega_r} = \sqrt{\frac{\tau_p^2 v_g a_N \eta_i}{q V_P}(I - I_T)\left(1 + \frac{\Gamma a_P}{a_N}\right)}. \tag{A4.17}$$

From these equations, we see that both the relaxation oscillation frequency and the relative damping increase with the square root of the injection current above threshold. For reasons that will be discussed further in Appendix 6.1, it is clear that biasing the laser as far above threshold as possible is desirable for high speed modulation.

References

Ackerman, E. I., Kasemset, D., Wanuga, S., Hogue, D. and Komiak, J. 1990. A high-gain directly modulated L-band microwave optical link, *Proc. IEEE MTT-S Int. Microwave Symp.*, paper C-3, 153–5.
Alferness, R., Korotky, S. and Marcatili, E. 1984. Velocity-matching techniques for integrated optic traveling wave switch/modulators, *IEEE J. Quantum Electron.*, **20**, 301–9.
Betts, G. E. 1989. Microwave bandpass modulators in lithium niobate, *Integrated and Guided Wave Optics*, 1989 Technical Digest Series, vol. 4, Washington, DC: Optical Society of America, 14–17.

Bode, H. W. 1945. *Network Analysis and Feedback Amplifier Design*, New York: Van Nostrand, Section 16.3.

Bridges, W., Sheehy, F. and Schaffner, J. 1991. Wave-coupled LiNbO$_3$ electrooptic modulator for microwave and millimeter-wave modulation, *IEEE Photon. Technol. Lett.*, **3**, 133–5.

Carlin, H. J. 1954. Gain-bandwidth limitations on equalizers and matching networks, *Proc. IRE*, **42**, 1676–85.

Coldren, L. A. and Corzine, S. W. 1995. *Diode Lasers and Photonic Integrated Circuits*, New York: John Wiley & Sons, Chapter 2.

Cox, C. H., III 1986. Unpublished laboratory notes.

Cox, C. H., III and Ackerman, E. I. 1999. Limits on the performance of analog optical links. In *Review of Radio Science 1996–1999*, ed. W. Ross Stone, Oxford: Oxford University Press, Chapter 10.

Dolfi, D. and Ranganath, T. 1992. 50 GHz velocity-matched broad wavelength lithium niobate modulator with multimode active section, *Electron. Lett.*, **28**, 1197–98.

Fano, R. M. 1950. Theoretical limitations on the broadband matching of arbitrary impedances, *J. Franklin Inst.*, **249**, 57–83; **249**, 139–54.

Georges, J., Kiang, M., Heppell, K., Sayed, M. and Lau, K. 1994. Optical transmission of narrow-band millimeter-wave signals by resonant modulation of monolithic semiconductor lasers, *IEEE Photon. Technol. Lett.*, **6**, 568–70.

Goldsmith, C. L. and Kanack, B. 1993. Broad-band reactive matching of high-speed directly modulated laser diodes, *IEEE Microwave and Guided Wave Letters*, **3**, 336–8.

Gopalakrishnan, G., Bulmer, C., Burns, W., McElhanon, R. and Greenblatt, A. 1992a. 40 GHz, low half-wave voltage Ti:LiNbO$_3$ intensity modulator, *Electron. Lett.*, **28**, 826–7.

Gopalakrishnan, G., Burns, W. and Bulmer, C. 1992b. Electrical loss mechanisms in travelling wave LiNbO$_3$ optical modulators, *Electron. Lett.*, **27**, 207–9.

Gulick, J. J., de La Chapelle, M. and Hsu, H. P. 1986. Fundamental gain/bandwidth limitations in high frequency fiber-optic links, *High Frequency Optical Communications*, *Proc. SPIE*, **716**, 76–81.

Izutsu, M. 1996. Band operated light modulators, *Proc. 25 General Assembly of the International Union of Radio Science*, Lille, France, August 28–September 5, 1996, paper DC-4, 639.

Kato, K., Hata, S., Kawano, K., Yoshida, H. and Kozen, A. 1992. A high-efficiency 50 GHz InGaAs multimode waveguide photodetector, *IEEE J. Quantum Electron.*, **28**, 2728–35.

Kato, K., Kozen, A., Maramoto, Y., Nagatsuma, T. and Yaita, M. 1994. 110-GHz, 50%-efficiency mushroom-mesa waveguide p-i-n photodiode for a 1.55-μm wavelength, *IEEE Photon. Technol. Lett.*, **6**, 719–21.

Lee, H. 1998. Personal communication.

Noguchi, K., Miyazawa, H. and Mitomi, O. 1994. 75 GHz broadband Ti:LiNbO$_3$ optical modulator with ridge structure, *Electron. Lett.*, **30**, 949–51.

Onnegren, J. and Alping, A. 1995. Reactive matching of microwave fiber-optic links, *Proc. MIOP-95*, Sindelfingen, Germany, 458–62

Pozar, D. M. 1993. *Microwave Engineering*, Boston: Addison-Wesley, 325–7.

Prince, J. L. 1998. Personal communication.

Roberge, J. K. 1975. *Operational Amplifiers Theory and Practice*, New York: John Wiley & Sons, 95.

Van Valkenburg, M. E. 1964. *Network Analysis*, 2nd edition, Englewood Cliffs, NJ: Prentice-Hall, Inc., 338–9.

Wang, W., Tavlykaev, R. and Ramaswamy, R. 1996. Bandpass traveling-wave modulator in LiNbO$_3$ with a domain reversal, *Proc. IEEE Lasers Electro-Opt. Soc. Annu. Meet. (LEOS'96)*, 99–100.

Weisser, S., Larkins, E., Czotscher, K., Benz, W., Daleiden, J., Esquivias, I., Fleissner, J., Ralston, J., Romero, B., Sah, R., Schonfelder, A. and Rosenzweig, J. 1996. Damping-limited modulation bandwidths up to 40 GHz in undoped short-cavity multiple-quantum-well lasers, *IEEE Photon. Technol. Lett.*, **8**, 608–10.

5

Noise in links

5.1 Introduction

In Chapters 2 through 4 we have shown how a single formalism can be used to describe the gain and frequency response of both direct and external modulation links. We continue with that same approach in this chapter. However, we will see that because different noise sources dominate in each type of link, the specific form of the link noise model depends on the type of link.

Up to this point all the signal sources we have dealt with were deterministic, in the sense that we could express their output voltage at any instant of time in terms of a known function of time, say $v(t)$. In the case of the noise sources discussed here, there are – at present – no known expressions for any of the noise sources that give the noise source output as a deterministic function of time. Consequently we are forced to use the next best description, which is to describe the noise source output in terms of its statistical properties.

There are many statistical descriptors that could be used; by far the most common one for describing noise sources in electrical and optical applications is the mean-square value. There are primarily two bases for the popularity of the mean-square value. One is that it can be derived from the statistical distribution for the noise source, without ever knowing the underlying deterministic function. The other reason is that the mean-square value corresponds to the heating effect generated by the noise source. This latter property permits the effects of noise sources to be compared directly to the effects of deterministic sources, such as signal sources.

With deterministic sources, the mean-square value of the voltage – or its square root, the root-mean-square (RMS) value – can be calculated from the deterministic function for the source:

$$v_{\text{rms}} = \sqrt{\frac{1}{T} \int_0^T v^2(t)\, \mathrm{d}t} \equiv \sqrt{\langle v^2(t) \rangle}, \qquad (5.1)$$

159

where T is some time interval over which the noise is a stationary random variable.

5.2 Noise models and measures

5.2.1 Noise sources

In this section we develop circuit representations for the dominant noise sources in a link: thermal, shot and relative intensity noise. Our treatment of the first two sources of noise is based on Motchenbacher and Fitchen (1973) and Pettai (1984); the development of laser relative intensity noise is based on Petermann (1988).

5.2.1.1 Thermal noise

In back-to-back papers in *Physical Review*, Johnson (1928) and Nyquist (1928) reported the experimental observation of, and the theoretical basis for, respectively, thermal noise. This noise arises from the thermally induced, random movement of electrical carriers in conductors. Although Nyquist's development can be summarized in a couple of pages (see for example Robinson, 1974), we choose not to repeat that development here but rather simply state the results. Nyquist showed that the mean-square value of the noise voltage $\langle v_t^2 \rangle$ across a resistor R at an absolute temperature T as measured in a small bandwidth Δf is given by

$$\langle v_t^2 \rangle = 4kTR\Delta f, \tag{5.2}$$

where k is Boltzmann's constant, which has a value of 1.38×10^{-23} J/K. The temperature used in calculating the noise is the physical temperature of the loss, which is usually the physical temperature of the component.[1]

Recall from (2.7) that the available power, $p_{s,a}$, is simply voltage squared divided by $4R$. Thus the available thermal noise power can be obtained by simply dividing (5.2) by $4R$; the result is

$$p_{t,a} = kT\Delta f. \tag{5.3}$$

If we assume $\Delta f = 1$ Hz and $T = 290$ K – common values in noise calculations – then $p_{t,a} \cong 4 \times 10^{-21}$ W or approximately -174 dBm.

An implication of the above result is that for the purposes of analyzing the circuit impacts of thermal noise, we can replace the noisy, physical resistors, as shown in Fig. 5.1(a), by noiseless resistors with the same resistance as the physical resistors

---

[1] For example (Bridges, 2000), the noise temperature associated with the radiation resistance of an antenna is the physical temperature of the sources from which it receives radiation, i.e. 3 K for black body radiation from space. On the other hand, the resistive losses in the antenna's conductors generate noise according to their (local) physical temperature.

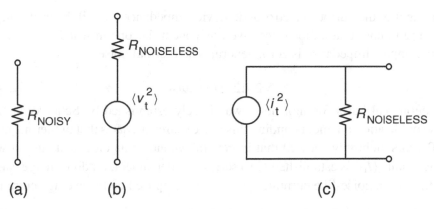

Figure 5.1 Circuit representation of a physical, noisy resistor (a) and two equivalent circuits for representing the thermal noise of the resistor: (b) voltage source in series with a noiseless resistor and (c) current source in parallel with a noiseless resistor.

that are connected in series with voltage sources whose mean-square values are given by (5.2); see Fig. 5.1(b).

From elementary circuit theory (see for example Van Valkenburg, 1964) we know that the circuit shown in Fig. 5.1(c) is equivalent to the circuit shown in Fig. 5.1(b), provided that the mean-square current is

$$\langle i_t^2 \rangle = 4kT\Delta f / R. \tag{5.4}$$

Representing the thermal noise as a current source is handy in the following, since both of the other noise sources discussed below are more naturally represented as current sources.

Nyquist also showed that thermal noise has a flat or "white" spectrum, over the range of frequencies

$$0 \le f \ll \frac{kT}{h}, \tag{5.5}$$

where h is the Planck constant, 6.6×10^{-34} J s. Continuing to use room temperature, the upper bound of (5.5) is about 6 THz. In practice the detection method and parasitic components always limit the observed bandwidth of thermal noise to significantly less than this value. Since the highest reported bandwidth for an electro-optic device is more than an order of magnitude less than this value, we assume that thermal noise is white throughout this book.

In practice one usually encounters complex rather than purely resistive impedances. For these situations the resistance in (5.2) is replaced by the real part of the complex impedance. Consequently thermal noise shows up not only in all the resistors used in passive matching and filtering circuits but also in the resistive component of the various electro-optic device impedances. It will be useful in the following to note that an ideal reactive component, i.e. one in which the real part of the complex impedance is zero, generates no thermal noise.

5.2.1.2 Shot noise

The theoretical basis for shot noise was largely established by Schottky (1918), whose publication on understanding this noise source predates that of thermal noise by 10 years. Schottky showed that in general whenever an electrical current with average value $\langle I_D \rangle$ is established via a series of independent, random charge carrier transits, then a noise-like current, $i_{sn}(t)$, is superimposed on the average current:

$$i_D(t) = \langle I_D \rangle + i_{sn}(t), \tag{5.6}$$

where $\langle i_{sn}(t) \rangle = 0$. Further, he showed that the mean-square value of the shot noise current over an interval τ, $\langle i_{sn}^2 \rangle$, is simply related to the average current of the interval by a constant:

$$\langle i_{sn}^2 \rangle = \frac{|q|}{\tau} \langle I_D \rangle, \tag{5.7}$$

where q is the charge of the current carrier, usually electrons, which have a charge of -1.602×10^{-19} C.

The photodetector current in an optical link arises from a series of independent, random events – due to the statistical arrival of photons on the photodetector. Consequently there is a shot noise current superimposed on the average photodetector current.

As we just did with thermal noise, it is often more useful to express the shot noise in the frequency domain, so we can readily see the scaling effects of bandwidth. Taking the Fourier transform of the shot noise current, we can rewrite (5.7) over a bandwidth Δf:

$$\langle i_{sn}^2 \rangle = 2q \langle I_D \rangle \Delta f, \tag{5.8}$$

where the factor of 2 in (5.8) comes from integrating the effects of both positive and negative frequencies.

Unlike in the thermal noise case, it is not possible to state, in general, the bandwidth over which the spectrum of shot noise is "white". The actual spectrum depends on the internal physics of the device exhibiting the noise. Consider the case with a photodiode. While the generation of shot noise within the space charge region of the photodiode is via independent, impulse-like current pulses, the photodiode output invariably involves drift of the carrier across the space charge region. This in turn broadens the pulse in time or correspondingly low pass filters the frequency response. However, for the purposes of this book, we assume that these transient time effects are negligible over the range of frequencies employed in the link and hence that the frequency spectrum of shot noise is "white", i.e. it is flat, like the spectrum of thermal noise.

From (5.8) it is clear that the shot noise decreases as the average current decreases. Consequently by decreasing the average current, we eventually reach the point where the shot noise is less than the thermal noise. The average current that produces a shot noise equal to the thermal noise, $I_{S=T}$, can be calculated by setting (5.4) equal to (5.8) and solving for the average current:

$$I_{S=T} = \frac{2kT}{qR},\tag{5.9}$$

which is approximately 1 mA when $R = 50\ \Omega$ and $T = 290$ K.

5.2.1.3 Relative intensity noise

There are many factors that contribute to the random fluctuations of a laser's output power, i.e. noise. Among the principal contributors to laser noise are the random emissions of spontaneous and stimulated photons in time, the selection at the laser facet of which photon is reflected and which is emitted and the fluctuations in the laser pump current. There were several early studies of this noise, see for example Lax (1960), Armstrong and Smith (1965) and McCumber (1966). More recently laser noise has been studied in considerable detail in a pair of papers by Yamamoto *et al.* (1983).

A common measure of the random fluctuations in a laser's optical output power is the relative intensity noise, *rin* (see for example Petermann, 1988). If we express the total optical power, $p_O(t)$, in an analogous form to current in the development of shot noise, viz. $p_O(t) = \langle P_O \rangle + p_{rin}(t)$ where $\langle p_{rin}(t) \rangle = 0$, then a definition of *rin* that is useful in link design is

$$rin = \frac{\langle p_{rin}^2(t) \rangle}{\langle P_O \rangle^2} = \frac{2\langle p_{rin}^2(t) \rangle}{\langle P_O \rangle^2 \Delta f},\tag{5.10}$$

where again the factor of 2 in the frequency domain expression comes from integrating the effects of both positive and negative frequencies.

The first place we see the effects of *rin* in a link is at the photodiode's electrical output, where the optical power has been converted to an electrical current. Since both the optical powers in (5.10) are squared, the photodiode responsivity cancels out and we can write (5.10) directly in terms of photodiode currents:

$$RIN = 10\log\left(\frac{2\langle i_{rin}^2(t) \rangle}{\langle I_D \rangle^2 \Delta f}\right),\tag{5.11}$$

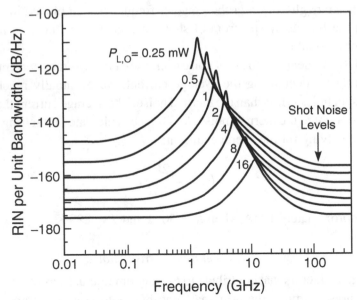

Figure 5.2 Plot of the calculated RIN vs. frequency for an in-plane diode laser with the laser's average optical power as a parameter (Coldren and Corzine, 1995, Fig. 5.18. © 1995 John Wiley & Sons, Inc., this material is used by permission of John Wiley & Sons, Inc.)

where $i_{\text{rin}} = i_{\text{o,d}}$ when there is no intentional intensity modulation of the optical carrier. In (5.11) we have expressed the RIN in dB/Hz, which is common. Also from (5.11) we see that the *rin* noise source can be represented in a circuit as a current source, like the thermal and shot noise sources.

In link design we are often given the laser RIN and need to calculate the noise current that results. This is easily obtained from (5.11):

$$\langle i_{\text{rin}}^2 \rangle = \frac{\langle I_{\text{D}} \rangle^2}{2} 10^{\frac{RIN}{10}} \Delta f. \tag{5.12}$$

However, unlike these other two noise sources, the RIN frequency spectrum is not flat over all the frequencies of interest in links. Figure 5.2 is a plot from Coldren and Corzine (1995) that shows the calculated RIN spectrum for a typical in-plane diode laser. The RIN spectrum is constant at low frequencies, peaks at the relaxation resonance frequency and then falls to the shot noise level above resonance.

In the RIN dominated portion of the spectrum, i.e. less than the relaxation peak, the noise power should vary as the square of the laser power, whereas above this peak the noise power should vary linearly with the laser power (Bridges, 2000). Indeed over the range of optical powers from 0.25 mW to 4 mW we would expect a 24 dB change in the RIN noise power but only a 12 dB change in the shot noise power. From Fig. 5.2 we see 22 dB and 11 dB respectively.

The RIN spectrum of a solid state laser has the same general features as the RIN spectrum of a diode laser. However, because the upper state lifetime is orders of magnitude longer than it is in diode lasers, the relaxation resonance (recall (4.3)) for diode-pumped solid state lasers is typically 1 MHz or less.

The above discussion of RIN is strictly valid only for single mode lasers. However, several lasers commonly encountered in link design are multi-mode; among them most solid state lasers and Fabry–Perot diode lasers. The number of longitudinal modes in the laser cavity has two primary effects on the intensity noise of a laser: mode partition noise and peaks in the RIN spectrum.

Mode partition noise contributes to the intensity noise of individual modes in a multi-mode laser, which is higher than the single mode RIN (see for example McCumber, 1966; Yamada, 1986). It turns out, however, that the noise associated with the sum of the powers from all the modes of a multi-mode laser closely approximates the single mode RIN. That is, we do not see mode partition noise unless we "partition" away some of the modes. In link design one rarely encounters the case where the output of a multi-mode laser has been optically filtered to provide a single mode output. Consequently we neglect mode partition noise in this book and treat the sum of the noise associated with all the modes collectively as single mode RIN.

The other primary noise impact of multi-mode lasers is on the output spectrum. Each of the longitudinal modes is actually at a slightly different frequency. Since the equations describing laser operation contain non-linear terms, there is beating among the multiple modes. As we will see in Chapter 6, a second-order non-linearity generates new frequencies at the sum and difference of the original frequencies. The sum of two optical frequencies is clearly an even higher frequency, but the difference between two closely spaced high frequency modes can be a very low frequency; sufficiently low in fact that it can occur within the link bandwidth. In a typical diode-pumped, solid state laser, the frequency difference between two adjacent longitudinal modes is about 4 GHz. A plot of an experimentally measured spectrum for such a laser is shown in Fig. 5.3. The relaxation resonance for this laser is so low, about 200 kHz, that it does not show up on this plot. However, we do see peaks in the spectrum about every 4 GHz, which corresponds to the intermodal frequency spacing for this laser. While these peaks are not really "noise" in the sense that they originate from a random process, they can certainly be classed as "interference" if they appear in the passband of the desired signal (Bridges, 2000).

The same effect occurs among the multiple modes of a Fabry–Perot diode laser. To get a feel for the frequencies of a diode laser's intermodal peaks, recall the example following (2.4) that calculates a typical diode laser modal spacing to be 0.91 nm, which corresponds to about 162 GHz. Thus we do not typically see

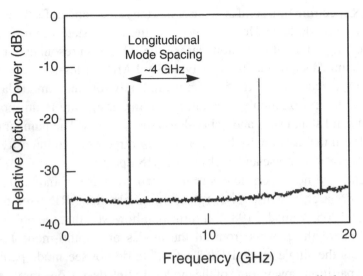

Figure 5.3 Plot of experimentally measured RIN spectrum from a diode-pumped, solid state laser with multiple longitudinal modes, showing peaks in the spectrum at the intermodal frequencies.

intermodal peaks when using diode lasers because their first intermodal peak occurs at a frequency that is usually above the passband of most links.

The RIN of a laser is often quoted as a single number. As Figs. 5.2 and 5.3 make clear, this may be acceptable if the bandwidth of the link spans a frequency range where the RIN spectrum is relatively independent of frequency.

The alert reader may ask: if the photodetector output is viewed over a frequency span where the RIN is relatively flat, how does one tell the difference between the RIN and shot noise? One method emerges from a comparison of (5.8) and (5.12): shot noise *power* increases linearly with average photodetector current, while RIN *power* increases as the square of the same current.

Conversely RIN decreases faster than shot noise as the average photodetector current decreases. Therefore for sufficiently low optical powers, the RIN is masked by the shot noise. A useful parameter is the minimum detectable RIN, which can be obtained by substituting (5.8) for $\langle i_{\mathrm{rin}}^2 \rangle$ in (5.11) to yield

$$RIN_{\substack{\mathrm{shot}\\\mathrm{noise}}} = 10\log\left(\frac{2q}{\langle I_{\mathrm{D}}\rangle}\Delta f\right), \tag{5.13}$$

after using one power of the denominator current to cancel the numerator current. For example, evaluating (5.13) for an average current of 1 mA and in a 1 Hz bandwidth, yields a minimum detectable RIN over shot noise of about −155 dB.

5.2.2 Noise figure

Measuring the combined effect of noise sources such as those discussed in Section 5.2.1 is not a practice that is limited to analog links, but is common throughout the RF field. A standard measure of the effects of noise in a device or circuit is the noise factor, nf, (IRE, 1960) or the logarithm of the noise factor, the noise figure, NF, i.e. $NF = 10 \log_{10}(nf)$. The noise factor is defined as follows: the noise factor of a two-port device is the ratio of the available output noise power per unit bandwidth to the portion of that noise caused by the actual source connected to the input terminals of the device, measured at the standard temperature of 290 K.

When the input noise is simply the thermal noise from a matched resistive load, i.e. when $n_{in} = kT\Delta f$, then an equivalent and more useful definition of noise figure is the ratio of the input signal-to-noise ratio to the output signal-to-noise ratio at 290 K:

$$NF \equiv 10 \log \left(\frac{s_{in}/n_{in}}{s_{out}/n_{out}} \right). \tag{5.14}$$

This form of the noise figure definition makes clear that noise figure is a measure of the degradation of the signal-to-noise ratio caused by a particular device or network. Since the output signal-to-noise ratio is at best equal to the input signal-to-noise ratio, the minimum noise figure value, i.e. the best noise figure, is 0 dB.[2]

An alternative form of (5.14), which will be useful in the following, can be obtained by expressing the output signal and noise in terms of the gain and input signal and noise, viz.:

$$s_{out} = g_i s_{in}, \tag{5.15a}$$

$$n_{out} = g_i n_{in} + n_{add}, \tag{5.15b}$$

where n_{add} is the noise added by the device or circuit. Substituting (5.15a) and (5.15b) into (5.14) we obtain

$$NF = 10 \log \left(1 + \frac{n_{add}}{g_i n_{in}} \right). \tag{5.16}$$

The (5.16) expression for noise figure opens further insights into this parameter. One is that it confirms our intuitive argument that the minimum $NF = 0$ dB, which occurs when $n_{add} = 0$ W. A second insight is that the apparent dependence of NF on the signal is explicitly eliminated. A third insight is that NF measures the effects of noise added by a device or circuit by translating that noise to an equivalent

---

[2] If the temperature of the device or circuit is cooled below the reference temperature of 290 K, then one can obtain apparently negative noise figures. For this reason, the IEEE definition specifies a source temperature of 290 K. When sources or systems at other than 290 K are encountered, "noise temperature" rather than noise figure is a better parameter to use.

additional noise source at the input. This is accomplished by dividing the added noise by the gain between the input and the additional noise source.

Since all the noise sources we presented in Section 5.2.1 had a bandwidth dependence, it is natural to ask: what is the bandwidth dependence of noise figure? To answer this question, recall that both thermal and shot noise are "white", at least for any frequency typically encountered in link design. Further RIN is "white" also as long as the frequency band of interest does not include either the relaxation resonance or one of the intermodal beat frequencies. Under these conditions both n_{add} and n_{in} have the same bandwidth dependence. Consequently under these conditions the noise figure is independent of bandwidth. We adopt this point of view throughout this book.[3]

5.3 Link model with noise sources

5.3.1 *General link noise model*

We now have the tools we need to form a noise model of an optical link and evaluate the effects of the noise sources. Thus we could take a link model like the one shown in Fig. 4.27(d), replace all the resistors by a series connection of a thermal noise source and a noiseless resistor, add the RIN and shot noise current sources in parallel with the photodiode output and calculate the noise figure using (5.16). However, the complexity of this approach tends to obscure some important points. Instead we build up to the general case by considering two cases of practical interest – one where the laser RIN and one where the shot noise is the dominant noise source at the photodetector. We defer to Section 5.5.2 the case where the dominant noise source is the thermal noise generated by the real part of the photodetector load impedance.

In presenting the definitions of the various types of noise, we have considered only one noise source at a time. Clearly in the following we will be trying to evaluate the combined effects of multiple noise sources on the link noise figure. Thus the question arises: how do we combine multiple noise sources? The answer is simple: we add their mean-square voltages or currents, just as we would add two sinusoidal sources of different frequencies. The only qualification on the previous statement is that the noise sources must be uncorrelated, i.e. that the voltage or current from each noise source is independent of the voltage or current of the other noise sources.

From Section 5.2.2 we saw that the link gain plays a central role in the calculation of noise figure. In Chapter 4 we saw that when the link contains complex

---

[3] When these conditions are not true, then the noise figure has a frequency, and hence a bandwidth, dependence. In such cases the term "spot noise figure" is used, and commonly denoted $NF(f)$. Circuit elements (R, L, C) as well as device characteristics can (and usually do) make NF a function of frequency.

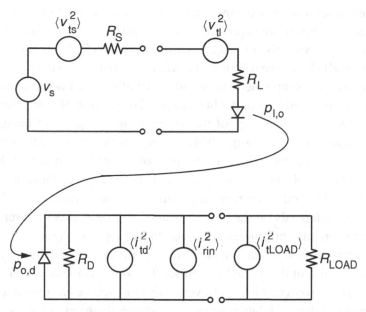

Figure 5.4 Schematic showing the noise sources of a directly modulated link under the assumption that laser RIN dominates over the photodetector shot noise.

impedances, the corresponding link gain is also complex, i.e. it has both a magnitude and a phase. In such cases, it is the magnitude of the gain that enters into the noise figure calculation. Since many of the points about noise figure that we intend to make below do not depend on the frequency response of the link, we begin by considering links with only real impedances, i.e. resistors, and thus we can use the corresponding gain expressions we derived in Chapter 3. This assumption continues in force until the last section, 5.5.3.2, where the investigation of the impacts of mis-match requires us to include complex impedances and thereby requires us to use the magnitude of the complex gain.

5.3.2 RIN-dominated link

The noise figure of direct modulation links is often dominated by the diode laser RIN. Figure 5.4 is a circuit showing the thermal noise sources at the modulation end of the link and the RIN noise source at the detection end of a directly modulated link. The thermal noise associated with the diode laser comes from the resistive component of the diode laser impedance.

To focus the discussion on the noise properties of this link, we are using a simple magnitude impedance match between the modulation source and the diode laser: $R_S = R_L$. Further we assume that the load resistance is equal to the source resistance.

There are several paths we can take to arrive at the noise figure of this circuit. One way is to determine an expression for n_{add} and then use (5.16) to calculate the noise figure. To pursue this approach, we first note from (5.15b) that n_{add} represents the effects of all the internal link noise sources at the link output. Consequently to calculate n_{add} we simply need to calculate the effects that noise sources located throughout the link produce at the link output. The process of doing this is greatly simplified if we take advantage of the superposition property of linear networks (see for example Van Valkenburg, 1964). Superposition states that the response of a network with multiple independent sources is equivalent to the sum of the network responses to each of the independent sources individually, i.e. with only one source operating and all the other sources set to zero. The effect on the network of setting a source to zero depends on the type of source; voltage sources are replaced with short circuits and current sources are replaced with open circuits.

For the purposes of this example we consider only two contributions to n_{add}. One is the thermal noise of the diode laser. Since this source is located in the same loop as the modulation source, it is easy to verify that this source has the same gain from input to output as the modulation source. Consequently the effect of the diode laser thermal noise source at the link output – and therefore its power contribution to n_{add} – is simply $g_i kT$.

The other contribution to n_{add} is the RIN source. Since it is located at the link output its contribution to n_{add} can be written by inspection to be a noise power of $\langle i_{rin}^2 \rangle R_{LOAD}$.

In practice there are additional noise sources at the photodiode end of the link, two of which are shown in Fig. 5.4: the thermal noises from the photodiode resistance and the load resistance. In many analog link designs, the following example among them, the RIN and/or the photodiode current are sufficiently high that the RIN dominates over the thermal noise sources.

The remaining noise power we need to evaluate (5.16) is n_{in}, which is just the thermal noise power of the source resistor; hence $n_{in} = kT$.

Substituting these three terms into (5.16) yields an expression for the noise figure of a RIN-dominated, directly modulated link with passive magnitude impedance matching:

$$NF = 10\log\left(1 + \frac{g_i kT + \langle i_{rin}^2 \rangle R_{LOAD}}{g_i kT}\right) = 10\log\left(2 + \frac{\langle i_{rin}^2 \rangle R_{LOAD}}{s_t^2 r_d^2 kT}\right). \quad (5.17)$$

In the second expression for the noise figure we have substituted the link gain from (3.7).

Figure 5.5 is a plot of (5.17) vs. RIN using the nominal set of link parameters listed in Table 5.1. Recall that the gain of a directly modulated link is independent

Table 5.1 *Parameter values used in the plot
of (5.17) in Fig. 5.5*

Parameter	Value
Laser slope efficiency (W/A)	0.2
Photodiode responsivity (A/W)	0.8
Photodiode load resistance (Ω)	50
Average photodiode current (mA)	1
Temperature (K)	290

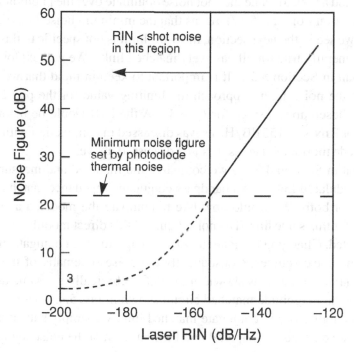

Figure 5.5 Plot of the calculated noise figure vs. RIN for the circuit shown in Fig. 5.4.

of the average optical power; consequently we can choose a nominal value for the average photodiode current and keep it fixed for these calculations.

The results confirm our intuition that as the RIN increases so does the noise figure. Also plotted in Fig. 5.5 is the contribution to the link noise figure made by a thermal noise source at the photodiode end of the link. As we see from the link schematic, Fig. 5.4, this noise source is connected in parallel with the RIN noise source. Hence we can readily calculate the noise figure due to this noise source by substituting (5.4) for the RIN noise current in (5.17). The resulting noise figure, ~22 dB, is plotted as a horizontal line in Fig. 5.5, since thermal noise is independent of the RIN. We see that for the particular set of parameters assumed for this example, the noise figure goes from being RIN-limited to photodiode-thermal-noise-limited, without ever being shot-noise-limited. This example points out that it is often important to check the range of validity for assumptions, such as assuming that the RIN dominates the photodiode noise sources. In this particular example we see that obtaining a laser with a lower RIN, such as an optically isolated DFB with a RIN of ~ -165 dB/Hz, would not reduce the noise figure.

What may not be as obvious at first is that (5.17) also predicts that when the RIN becomes negligible, the noise figure does not approach zero – assuming of course that we can find a way to make the shot noise dominate over the photodiode thermal noise. The constant of 2 in (5.17) means that the minimum noise figure of this link is 3 dB. As we see in the next section, the 3-dB limit is not specific to this particular link but is generally true for all passively matched links. We will explore this limit in some detail in Section 5.5.1. It is important to keep in mind that we would not actually see the noise figure approach the limiting value for the particular set of parameters chosen above, since for $I_D = 1$ mA the RIN would be below both the shot noise for RIN < -155 dB/Hz as was discussed in conjunction with (5.13) and the photodiode thermal noise, as was just discussed above.

Recall that in Section 4.3.2.3 we compared the modeled and measured gain of a directly modulated link with a bandpass conjugate impedance match at the laser, the detector or both. We would now like to compare the modeled and measured noise figure for this same link. The noise figure of this direct modulation link is also RIN-dominated. Changing the matching from magnitude to conjugate clearly does not affect the noise sources. Consequently the noise schematic of the Onnegren and Alping (1995) circuit, as shown in Fig. 5.6, is basically the same as the noise schematic of the magnitude matched circuit shown in Fig. 5.4. Therefore to calculate the noise figure of the conjugate matched link, we simply substitute the gain expressions we derived in Chapter 4 into the first form of the noise figure equation, (5.17).

The results of substituting (4.35) through (4.38) into (5.17) are plotted in Fig. 5.7. It is encouraging that this extremely simple model gives such relatively good agreement with the measured results. The results can be grouped into two categories: those that improved the noise figure and those that did not. The key to distinguishing between these two categories is the location of the impedance matching. Matching at the laser improves the noise figure while matching at the photodiode does not, whether done individually or in combination with matching at the laser. This is in contrast to the effects of impedance matching on gain, where matching at either the laser or the photodiode individually improved the gain, and when both were matched the gain improvement was the sum of the individual gain improvements. We present the basis for this in Section 5.4.

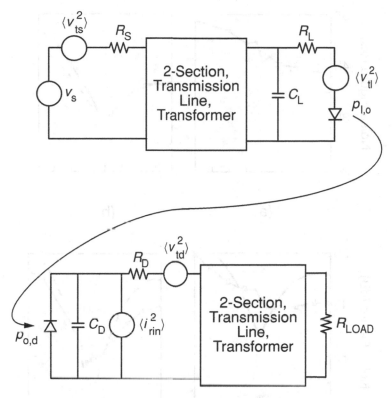

Figure 5.6 Noise equivalent circuit of the simplified version of the directly modulated link with bandpass matching (see also Fig. 4.20) (Onnegren and Alping, 1995, reprinted with permission from the authors).

5.3.3 *Shot-noise-dominated link*

In Section 5.2.1.3 it was pointed out that the relaxation peak of a diode laser is typically at least 5 GHz, whereas this peak for a solid state laser is typically less than 1 MHz. Thus the passbands of most links are below the relaxation resonance frequency of diode lasers but above this frequency for solid state lasers. Consequently the RIN of solid state lasers in the link passband is typically less than -175 dB/Hz, which is substantially less than the RIN of diode lasers, which is usually no better than -155 dB/Hz. By use of (5.13), we can neglect the RIN in external modulation links where a solid state laser is used as the CW source, for photodetector currents less than 100 mA! Thus in all such links the dominant source of noise at the photodiode is shot noise.

Figure 5.8 is a schematic showing the noise sources of a shot-noise-dominated, external modulation link using a Mach–Zehnder modulator biased at its most linear region, quadrature. As with the direct modulation link, we assume a magnitude impedance match between source and modulator for simplicity.

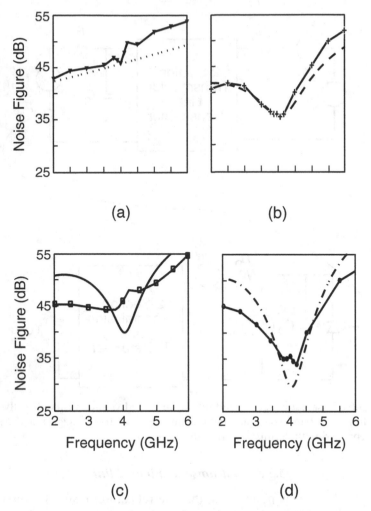

Figure 5.7 Plot of the modeled and measured link noise figure vs. frequency with (a) no impedance matching at either laser or detector, (b) laser impedance matching only, (c) photodiode impedance matching only and (d) matching at both laser and detector (Data from Onnegren and Alping, 1995. Reprinted with permission from the authors.)

Although the modulation device has changed from a diode laser to an external modulator, the modulator also has a resistive component to its impedance – which can be for example a matched termination. Thus the noise sources at the modulation end of this link are the same as those for the direct modulation link. At the photodiode end of the link, the shot noise current source replaces the RIN current source of the previous case.

Since the general form of the noise schematic for the shot-noise-dominated, external modulation link is identical to that of the RIN-dominated, direct modulation

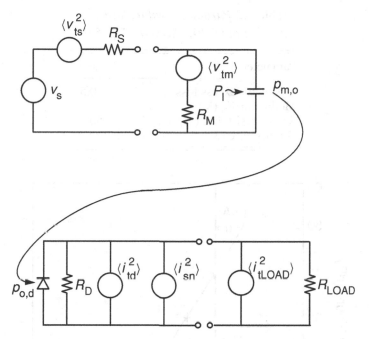

Figure 5.8 Schematic showing the noise sources for a shot-noise-dominated, externally modulated link with magnitude impedance matching.

link – if we continue to assume that the shot noise dominates over the photodiode thermal noise – we can substitute shot noise for RIN in (5.17) and immediately write the expression for the noise figure of this external modulation link:

$$NF = 10 \log \left(2 + \frac{\langle i_{sn}^2 \rangle R_{LOAD}}{g_i kT} \right).$$ (5.18)

Exploration of the dependence of noise figure on shot noise is a little more complex than our exploration of the noise figure dependence on RIN because the shot noise and the intrinsic gain have different dependences on the average photodetector current. We can make both of these dependences explicit by substituting the expressions for shot noise, (5.8), and intrinsic gain of an external modulation link with a Mach–Zehnder quadrature-biased modulator, (3.10), into (5.18):

$$NF = 10 \log \left(2 + \frac{2q I_D R_{LOAD}}{\left(\dfrac{T_{FF} P_I \pi R_S}{2 V_\pi} \right)^2 r_d^2 kT} \right).$$ (5.19)

We can simplify (5.19) a little by recognizing that $I_D = P_I r_d / 2$ when the optical loss between the modulator output and the photodiode can be neglected, i.e.

Table 5.2 *Parameter values used in the plot of (5.20) shown in Fig. 5.9*

Parameter	Value
Modulator excess loss	0.5
Modulator V_π (V)	2
Source resistance (Ω)	50
Temperature (K)	290

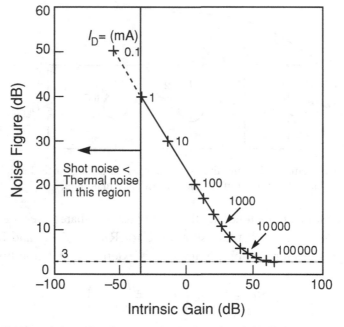

Figure 5.9 Plot of the noise figure vs. intrinsic gain of the link shown in Fig. 5.8 with average photodiode current as a parameter.

$T_{FF} = 1$. If we also continue to assume that $R_{LOAD} = R_S$, then making both these substitutions into (5.19) yields

$$NF = 10\log\left(2 + \left(\frac{2qV_\pi^2}{\pi^2 R_S kT}\right)\left(\frac{1}{I_D}\right)\right). \tag{5.20}$$

Figure 5.9 is a plot of (5.20) vs. (3.10), using the values for the other link constants listed in Table 5.2. Representative calculated values of the average photodiode current that yield a particular gain/noise figure pair are indicated by "+" symbols along the curve in Fig. 5.9. Keep in mind that the shot noise is directly proportional to the average photodiode current, which as we can see from the annotation in

Fig. 5.9, increases as we move down along the gain/noise figure curve. This means that unlike the direct modulation, RIN-dominated link discussed above, in the external modulation, shot-noise-dominated link the noise figure *decreases* as the shot noise *increases*.

The reason for what may at first seem to be paradoxical behavior arises from a combination of the properties of shot noise, the external modulation link gain, and the definition of noise figure. Recall that one way to view noise figure is as a measure of the effects of all the noise sources in a network referred to the input of the network. Applying this view to an optical link means that the effects of noise sources, which are physically located at the photodetector, are translated for the purposes of the noise figure calculation to equivalent noise sources at the link input. The transfer of a noise source from output to input is accomplished by dividing the output source by the intrinsic gain. Applying this notion to the case of the shot noise located at the photodiode means dividing a noise source whose noise power scales linearly with optical power by an intrinsic gain that scales quadratically with optical power. The result is an equivalent noise source at the link input whose noise power scales inversely with optical power! Thus the shot noise power is indeed increasing as the optical power increases, but the link gain is increasing faster than the shot noise power. Consequently the effect of the increased shot noise is a lower effective noise at the input, which in turn means a lower noise figure.

The other point to notice from Fig. 5.9 is the fact that once again the noise figure does not approach zero but 3 dB, even if we could have impracticably high photodiode currents and negligible RIN. However, as we calculated above for presently available solid state lasers, we may no longer be able to neglect the RIN at photodetector currents above 100 mA. Conversely, for low detector currents, the intrinsic gain decreases and the noise figure increases dramatically.

Recall that in Section 4.3.3.3 we compared the modeled and measured gain of an externally modulated link with a bandpass conjugate impedance match at the modulator, the detector or both (Prince, 1998). We would now like to compare the modeled and measured noise figure for this same link. The noise figure of this external modulation link is also shot noise dominated. Changing the matching from magnitude to conjugate clearly does not affect the noise sources. Consequently the noise schematic of the link with conjugate impedance match is basically the same as the noise schematic of the magnitude matched circuit shown in Fig. 5.8. Therefore to calculate the noise figure of the conjugate matched link, we simply substitute the gain expressions we derived in Chapter 4 into the first form of the noise figure equation, (5.17).

The results of substituting (4.52) through (4.55) into (5.17) are plotted in Fig. 5.10. As was the case with the direct modulation example, Section 5.3.2,

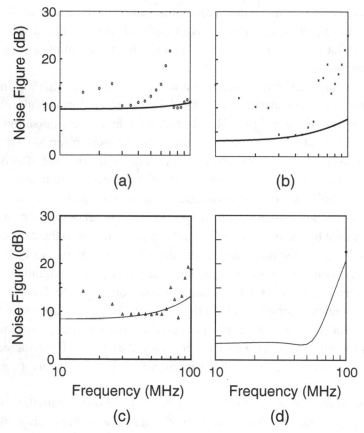

Figure 5.10 Plot of the modeled (solid lines) and measured (symbols) link noise figure vs. frequency with (a) no impedance matching at either modulator or detector, (b) modulator impedance matching only, (c) photodiode impedance matching only and (d) matching at both modulator and detector (Prince, 1998 reprinted with permission from the author).

the results of this external modulation example can be grouped into the same two categories: those located at the modulator improved the noise figure while those located at the photodetector did not – at least experimentally. Comparing theory and experiment in this case brings up two facts. One is that it is more difficult to model noise figure than gain. The other is that device resonances, which produce the noticeable noise figure spikes in Fig. 5.10, do not produce noticeable gain spikes that are as evident in Fig. 4.28.

5.4 Scaling of noise figure

In the previous sections we have seen the effects on noise figure that various combinations of slope efficiency, impedance matching and average optical power can have on RIN- and shot-noise-dominated links. Often the link designer is faced with the task of reducing the noise figure of an initial link design. To assist in selecting the techniques to use in any given situation, we would now like to investigate systematically the effectiveness of each of these three parameters in reducing the noise figure in RIN- and shot-noise-limited links.

The development below is an expanded version of the approach originally presented by Cox (1992).

5.4.1 Impedance matching

If we compare the link noise figure expressions for the RIN- and shot-noise-dominated links, (5.17) and (5.18) respectively, we see that they are identical except for the particular type of noise source. Further, impedance matching does not involve distinguishing between these two noise sources. Therefore for the purposes of this section the distinction between these two noise sources is not important; whatever we conclude about the effectiveness of impedance matching holds equally well for both noise sources. Consequently in this section we use the variable i_n^2 to represent both RIN and shot noise.

Next we need to choose the type of impedance matching. As we saw in Chapter 4, a conjugate or tuned match only affects the frequency response and not the magnitude of the in-band match, yet it complicates the analysis considerably. Of the magnitude matches, the resistive type adds another source of noise, while the transformer match does not. Thus a transformer magnitude match offers us a way of exploring the effects of impedance matching in a simple way without including additional noise in the process.

Recall from (3.4) that the intrinsic gain can be expressed as the product of the modulation and photodetection incremental modulation and detection efficiencies. If we compare the modulation efficiency of the transformer matched diode laser, (4.26), with the corresponding efficiency for a Mach–Zehnder modulator, (4.47), we see they have the same functional form. Since we are neglecting the frequency response terms in this discussion, we set $s = 0$ in the present application of both (4.26) and (4.46). We can now write a single expression for the incremental modulation efficiency of either type of modulation device as

$$\frac{p_{\mathrm{md},o}^2}{p_{\mathrm{s},a}} = \frac{s_{\mathrm{md}}^2}{R_{\mathrm{MD}}} = \frac{s_{\mathrm{md}}^2 N_{\mathrm{MD}}^2}{R_{\mathrm{S}}}, \tag{5.21}$$

where we have used the fact that the transformer turns ratio is chosen such that $R_{\mathrm{S}} = N_{\mathrm{MD}}^2 R_{\mathrm{MD}}$ to write the second form of (5.21).

The incremental detection efficiency for a transformer matched photodiode was given in (4.12); it is repeated below with $s = 0$ as we did with modulation devices:

$$\frac{p_{\text{load}}}{p_{\text{o,d}}{}^2} = r_{\text{d}}^2 N_{\text{D}}^2 R_{\text{LOAD}}. \tag{5.22}$$

The expression for intrinsic gain with transformer matching to both the modulation device and the photodiode is simply the product of (5.21) and (5.22)

$$g_{\text{i}} = s_{\text{md}}^2 N_{\text{MD}}^2 r_{\text{d}}^2 N_{\text{D}}^2, \tag{5.23}$$

where we have assumed that $R_{\text{S}} = R_{\text{LOAD}}$.

The transformer matching of the photodiode to the load also increases the load seen by the photodiode noise source from R_{LOAD} to $N_{\text{D}}^2 R_{\text{LOAD}}$. Making this substitution and (5.23) into the noise figure expression (5.17/18) yields

$$NF = 10 \log \left(2 + \frac{\langle i_{\text{n}}^2 \rangle N_{\text{D}}^2 R_{\text{LOAD}}}{s_{\text{md}}^2 N_{\text{MD}}^2 r_{\text{d}}^2 N_{\text{D}}^2} \right) = 10 \log \left(2 + \frac{\langle i_{\text{n}}^2 \rangle R_{\text{LOAD}}}{s_{\text{md}}^2 N_{\text{MD}}^2 r_{\text{d}}^2} \right), \tag{5.24}$$

where the second expression for the noise figure in (5.24) was obtained by canceling the photodiode turns ratio, N_{D}^2.

The implication of (5.24) is that impedance matching at the photodiode has no effect on the noise figure, regardless of whether the link is RIN- or shot-noise-dominated. Only matching to the modulation device can improve the link noise figure. And since (5.24) was derived in terms of a general modulation device, this conclusion holds for both direct and external modulation as well. We have seen experimental verification of these predictions in the RIN-dominated direct modulation and shot-noise-dominated external modulation noise figure results that were presented above.

As a consequence of the above result, we assume for the remainder of this chapter that the photodiode is simply connected to the load resistor without any further impedance matching circuitry.

5.4.2 Device slope efficiency

We would now like to explore the effects of modulation device slope efficiency and photodiode responsivity on the link noise figure. For this task we can continue to use the single designation of modulation device to represent both direct and external modulation. However, since the RIN and shot noise sources have a different functional dependence on the photodiode responsivity, we can no longer use a common designation for both types of noise.

Consider the shot-noise-dominated case first. The average photodiode current in the shot noise equation, (5.8), is equivalent to the product of the photodiode responsivity and the average optical power incident on the photodiode. Applying this equivalence to (5.8) yields an expression for the shot noise that shows the photodiode responsivity explicitly:

$$\langle i_{\text{sn}}^2 \rangle = 2q r_{\text{d}} \langle P_{\text{O,D}} \rangle. \tag{5.25}$$

Since our focus in this section is on slope efficiency and responsivity, we set both the turns ratio terms in (5.23) to one. Substituting (5.25) and this simplified form of (5.23) into the equation for shot-noise-dominated noise figure, (5.18), we obtain

$$NF = 10 \log \left(2 + \frac{2q r_{\text{d}} \langle P_{\text{O,D}} \rangle R_{\text{LOAD}}}{s_{\text{md}}^2 r_{\text{d}}^2 kT} \right) = 10 \log \left(2 + \frac{2q \langle P_{\text{O,D}} \rangle R_{\text{LOAD}}}{s_{\text{md}}^2 r_{\text{d}} kT} \right), \tag{5.26}$$

where the second form of the noise figure expression in (5.26) was obtained by canceling the numerator photodetector responsivity with one of the powers of the same responsivity in the denominator.

The second form of (5.26) makes clear that both the photodiode responsivity and the modulation device slope efficiency can affect the noise figure. However, because the modulation device slope efficiency appears as the second power and the photodiode responsivity only appears to the first power, the former is much more effective than the latter at improving the noise figure of a shot-noise-dominated link.

Now let us consider the RIN-dominated case. As we did in the shot noise case, we need to form an expression for the RIN that shows explicitly the dependence on photodiode responsivity. To do this we simply start with the expression for RIN current, (5.12), then express the average detector current in terms of the photodiode responsivity and the average optical power falling on the photodiode as we did for the shot noise, viz.:

$$\langle i_{\text{rin}}^2 \rangle = 10^{\frac{RIN}{10}} \frac{r_{\text{d}}^2 \langle P_{\text{O,D}} \rangle^2}{2}. \tag{5.27}$$

Substituting (5.27) and the simplified form of (5.23) into the RIN-dominated noise figure expression, (5.17), yields an equivalent form of this expression that shows the slope efficiency dependence explicitly:

$$NF = 10 \log \left(2 + \frac{10^{\frac{RIN}{10}} r_{\text{d}}^2 \langle P_{\text{O,D}} \rangle^2 R_{\text{LOAD}}}{s_{\text{md}}^2 r_{\text{d}}^2 kT} \right)$$

$$= 10 \log \left(2 + \frac{10^{\frac{RIN}{10}} \langle P_{\text{O,D}} \rangle^2 R_{\text{LOAD}}}{s_{\text{md}}^2 kT} \right), \tag{5.28}$$

where again the second noise figure expression in (5.28) is obtained by canceling the common photodiode responsivity terms in the numerator and denominator.

Table 5.3 *Summary of the effectiveness of impedance matching and slope efficiency on reducing the noise figure of RIN- and shot-noise-dominated links*

	RIN-dominated	Shot-noise-dominated
Impedance match	Modulation device – yes Photodiode – no	Modulation device – yes Photodiode – no
Slope efficiency	Modulation device – yes, quadratically Photodiode – no	Modulation device – yes, quadratically Photodiode – yes, linearly

Since the photodiode responsivity cancels out of the final form of (5.28), we see that this responsivity has no effect on the noise figure of RIN-dominated links. Thus in contrast to the shot-noise-dominated link, in the RIN-dominated link only the modulation device slope efficiency can be used to improve the noise figure. However, in common with the shot-noise-dominated link, the preceding conclusion holds independently of whether we are considering a directly or externally modulated, RIN-dominated link.

The results of the preceding two sections are summarized in Table 5.3; the results are independent of whether direct or external modulation is used.

5.4.3 Average optical power

Up to this point we have not required the details of how a particular value of modulation device slope efficiency was obtained. This permitted us to use a common formalism to treat both direct and external modulation. To explore the dependence of noise figure on optical power, however, requires separate expressions for direct and external modulation, because as we saw in Section 3.3.1 external modulation intrinsic gain depends on average optical power whereas direct modulation gain does not. Consequently we have four cases to consider: direct and external modulation in the RIN- and shot-noise-dominated regimes.

First let us consider the two direct modulation cases. Since the direct modulation slope efficiency does not depend – at least ideally – on average optical power, we can use the RIN- and shot-noise-dominated noise figure equations, (5.28) and (5.26), we already have by simply replacing s_{md} by s_1. Thus the RIN-dominated, direct modulation noise figure increases as the square of average optical power while the shot-noise-dominated, direct modulation noise figure only increases linearly with average optical power. In either case this suggests that one wants to operate a direct modulation link at low average optical power for the best noise figure. However, recall that this is in direct conflict with obtaining the maximum bandwidth from a

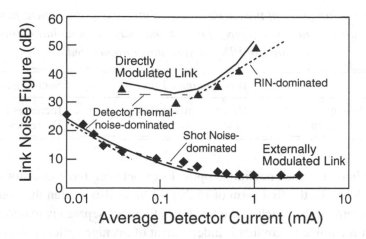

Figure 5.11 Plot of the noise figure vs. average optical power for a RIN-dominated, direct modulation and shot-noise-dominated external modulation link.

diode laser (see Section 4.2.1). This is but one example of the myriad tradeoffs that exist in link design. We explore a few more of these in Chapter 7.

Figure 5.11 is a plot of noise figure vs. average optical power. The data for the direct modulation case, which for sufficiently high optical powers was RIN-dominated, shows a noise figure that increases approximately as the square of average optical power.

For lower optical powers, the RIN noise current becomes less than the photodiode thermal noise current. Recall from the discussion surrounding Fig. 5.5 that when a directly modulated link becomes photodiode-thermal-noise-dominated, the resulting noise figure is independent of optical power. We have included this case in Fig. 5.11 by a horizontal dashed line. Admittedly with only two data points in this region we can only present a weak experimental case.

Next we consider the external modulation cases. To obtain an expression that shows explicitly the optical power dependence of noise figure for a RIN-dominated, external modulation link, we substitute the expression for the Mach–Zehnder slope efficiency, (2.18), into the first form of the RIN-dominated noise figure expression, (5.28) to obtain

$$NF = 10\log\left(2 + \frac{10^{\frac{RIN}{10}}\langle P_{\mathrm{O,D}}\rangle^2 R_{\mathrm{LOAD}}}{2\left(\dfrac{\pi R_{\mathrm{s}}}{2V_\pi}\right)^2 (T_{\mathrm{FF}}P_{\mathrm{I}})^2 kT}\right) = 10\log\left(2 + \frac{10^{\frac{RIN}{10}}R_{\mathrm{LOAD}}}{2\left(\dfrac{\pi R_{\mathrm{s}}}{2V_\pi}\right)^2 kT}\right),$$

$$(5.29)$$

Table 5.4 *Summary of the effectiveness of increasing the optical power on reducing the noise figure of direct and external modulation links that are RIN- or shot-noise-dominated*

	RIN-dominated	Shot-noise-dominated
Direct modulation	Increases quadratically	Increases linearly
External modulation	Independent	Decreases linearly

where we have assumed negligible optical loss between the modulator output and the photodiode. In the first form of (5.29) we have also written the optical power dependent term of the Mach–Zehnder slope efficiency separately to make it clearer. The result is a noise figure that is independent of average optical power.

Finally we consider the shot-noise-dominated, external modulation case. Using the same expression for the Mach–Zehnder slope efficiency, we now substitute (2.18) into the first form of (5.26), which is the expression for the shot-noise-dominated, external modulation noise figure:

$$NF = 10 \log \left(2 + \frac{2q \langle P_{\text{O,D}} \rangle R_{\text{LOAD}}}{\left(\dfrac{\pi R_{\text{s}}}{2V_\pi} \right)^2 (T_{\text{FF}} P_{\text{I}})^2 r_{\text{d}} kT} \right)$$

$$= 10 \log \left(2 + \frac{2q R_{\text{LOAD}}}{\left(\dfrac{\pi R_{\text{s}}}{2V_\pi} \right)^2 T_{\text{FF}} P_{\text{I}} r_{\text{d}} kT} \right). \qquad (5.30)$$

The second form of (5.30) shows formally what we noted earlier: that the noise figure of a shot-noise-dominated, external modulation link actually decreases linearly with increasing average optical power. Experimental data for this case are also plotted in Fig. 5.11. The data show the expected linear decrease in noise figure, at least over an intermediate range of average detector currents. For higher currents the decrease becomes sub-linear as the noise figure begins to approach asymptotically the 3-dB limit. For photocurrents less than the linear dependence range, the noise figure increases as the square of the photocurrent. The reason for this is straightforward: when the photodiode thermal noise dominates we have a noise power that is independent of optical power divided by a gain that is quadratically dependent on optical power.

Table 5.4 summarizes the effects of average optical power on the noise figure of RIN- and shot-noise-dominated, direct and external modulation links.

5.5 Limits on noise figure

In several of the preceding sections of this chapter we have noted that the limiting value of the link noise figure is not zero dB but some other value, usually 3 dB. We have also seen that as the gain drops below one, the noise figure increases in rough proportion to further decreases in gain. In this section we formally derive these and other limits on the noise figure.

In deriving the various limits, we use several terms and concepts that we need to define. One is a passive network. A network is passive if it contains no bias or signal power sources. A second term we need to define is lossless. A network is lossless if the real parts of all its impedances are zero. Combining these last two definitions, a passive, lossy network contains real, ohmic resistances that have thermal noise sources associated with them, but no other sources of bias or signal power.

A third concept is that of a perfect match between two networks. Recall from Section 4.4.1 that we introduced the concept of reflection coefficient, (4.56), as a measure of the match between two networks; in the case of that section the two networks were the modulation source and device. As we saw from (4.56) and Fig. 4.29, a perfect match exists when the reflection coefficient is zero.

5.5.1 Lossless passive match limit

We begin by assuming we have a link in which the matching network between the source and the modulation device is passive, lossless and achieves a perfect match. In this case we can use the noise figure equations we have already derived, (5.17) and (5.18), as the basis for establishing the limits on noise figure. Since we are investigating the noise figure limits, we need to include the thermal noise of the photodetector load resistance. We can represent this noise as a current source, (5.4), and locate it in parallel with the other two photodiode noise sources by simply adjusting the thermal source's magnitude by the turns ratio of the photodiode matching transformer, N_D^2:

$$\langle i_t^2 \rangle = \frac{1}{N_D^2} \frac{kT}{R_{\text{LOAD}}}.$$ (5.31)

We can now write a noise figure expression for a link in which no one of the photodiode noise sources is assumed to dominate by combining (5.17), (5.18) and (5.31):

$$NF = 10 \log \left(2 + \frac{N_D^2 R_{\text{LOAD}}}{g_i kT} \left(i_{\text{rin}}^2 + i_{\text{sn}}^2 + i_t^2 \right) \right).$$ (5.32)

Taking the limit of (5.32) as the intrinsic gain becomes very large we obtain

$$\lim_{g_i \to \infty} NF = 10 \log(2) = 3\,\text{dB}. \tag{5.33}$$

The fact that the limit is independent of intrinsic gain implies that this limit arises from noise sources at the input end of the link, since these are the only sources that do not involve a gain-dependent factor to translate them to the input.

We made very few assumptions in deriving (5.33): we did not assume a modulation method – direct or external; we did not assume any particular type of modulation device; nor did we assume that any particular photodiode noise source was dominant – RIN, shot or thermal noise. All we assumed were three conditions: that the modulation device was passively, losslessly and perfectly matched to the modulation source. Consequently we refer to the limit in (5.33) as the *lossless passive match limit*. To gain further insight into this limit, we now explore briefly the role that each of the three conditions just listed contributes to this limit.

As was pointed out just above, we only need to consider the input end of a general link, as shown in Fig. 5.12(a), in which we have a passive, lossless matching network connected between the modulation source and the modulation device. In terms of noise sources, a passive matching network can at most contain only thermal noise sources. If the passive network is also lossless, this implies that all the impedances within the network are purely reactive. Therefore there are no ohmic resistances – and thus no thermal noise sources – in a passive, lossless network. So for a noise analysis we can eliminate the matching circuit, since it does not add any noise.

Applying this result to the circuit shown in Fig. 5.12(a) permits us to simplify the link input to the form shown in Fig. 5.12(b). At this point we need the third condition for the passive match limit: the match. This condition constrains $R_S = R_{MD}$, which means that each noise source contributes equally to the noise at the link input. Applying the noise figure definition to the circuit in Fig. 5.12(b) yields a noise figure of 3 dB.

Removing any one of the three conditions for the lossless passive match limit invalidates the limit. For example, it is well known that if we allow an active circuit to do the matching – even if we constrain it to be lossless and to achieve a perfect match, we can obtain a noise figure less than 3 dB; commercial low noise amplifiers are presently available with noise figures below 1 dB. So the question is: why can active networks avoid the passive match limit? The short answer is that the input impedance of an active network can be resistive without being ohmic. Thus the input to an active network can match the source resistance without adding thermal noise. If the active network could achieve this condition without adding any other source of noise, then the noise figure of this ideal active network would be 0 dB. In practice actual active networks do have some noise; thus their noise figure can be

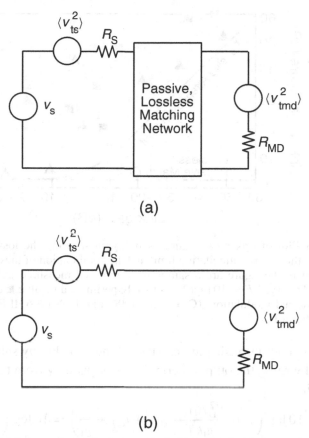

(a)

(b)

Figure 5.12 (a) Schematic for the input end of a link with lossless, passive matching between modulation source and modulation device. (b) Simplified version of (a) with the lossless matching circuit removed.

less than 3 dB but not 0 dB. A more rigorous discussion of this point is presented in Appendix 5.1.

Similarly, if we remove either (or both) the lossless and perfect match conditions, the lossless passive match limit no longer applies. We address the effects of relaxing these two conditions in Section 5.5.3.

5.5.2 *Passive attenuation limit*

We continue to assume that there is a passive, lossless and perfect match between the modulation source and the modulation device. However, in this section we wish to examine the limit on noise figure when the intrinsic gain is very low, i.e. when the link has substantial RF loss.

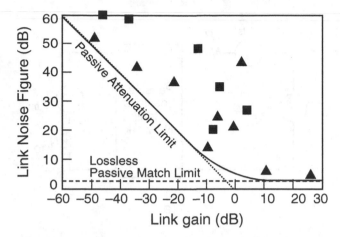

Figure 5.13 Plot of link noise figure vs. link gain showing the lossless passive
match limit, the passive attenuation limit and the combination of these two limits.
Also plotted in the figure are a sampling of reported measured gain and noise
figures. — Theory: $NF = 10 \log(2 + 1/g_i)$. Experimental results: ■ direct modulation; ▲ external modulation. (Cox *et al.*, 1996, Fig. 1. © 1996, IEEE, reprinted
with permission.)

The limit for small intrinsic gain can be obtained quickly by substituting (5.31)
into (5.32) and writing the output thermal noise separately from the RIN and shot
noise terms, viz.:

$$\lim_{\substack{g_i \to 0 \\ i_{\text{rin}}^2 + i_{\text{sn}}^2 \to 0}} NF = 10 \log\left(2 + \frac{N_D^2 R_{\text{LOAD}}}{g_i kT}\left(i_{\text{rin}}^2 + i_{\text{sn}}^2\right) + \frac{1}{g_i}\right) = 10 \log\left(\frac{1}{g_i}\right). \quad (5.34)$$

If we assume that the RIN and shot noises can be made negligible compared to
the thermal noise of the photodiode load, then the output thermal noise dominates
over the input thermal noise. The link noise figure limit under these conditions is
simply the reciprocal of the gain; in other words the noise figure approaches a value
equal to the RF loss of the link. Recall that the noise figure of a passive, matched RF
attenuator whose temperature is equal to the noise figure reference temperature is
simply equal to its attenuation (see for example Pettai, 1984, Section 9.4). Therefore
by analogy with this case we refer to the link noise figure limit under small gain as
the *passive attenuation limit*.

In contrast to the lossless passive match limit, the passive attenuation limit arises
from noise at the output end of the link. Since referring noises at the output back
to the input requires dividing by the gain, the passive attenuation limit is directly
dependent on gain.

Both the lossless passive match limit and the passive attenuation limit are plotted
in Fig. 5.13. The figure also contains a plot of the sum of these two limits. This

combined limit is easily expressed analytically from (5.34) by continuing to assume that the RIN and shot noise terms are negligible:

$$\lim_{i_{\text{rin}}^2 + i_{\text{sn}}^2 \to 0} NF = 10 \log \left(2 + \frac{N_D^2 R_{\text{LOAD}}}{g_i kT} \left(i_{\text{rin}}^2 + i_{\text{sn}}^2 \right) + \frac{1}{g_i} \right)$$

$$= 10 \log \left(2 + \frac{1}{g_i} \right). \tag{5.35}$$

Also plotted in Fig. 5.13 are data points of the measured gain and noise figure for a representative sample of experimental links that have been reported in the literature. It is reassuring to note that none of the experimental data points violate either of the limits presented above. However, none of the links were measured under more than one gain/noise figure condition. Consequently these data do not provide direct experimental confirmation of the detailed shape of the combination of these two limits presented in (5.35).

The simple form of the limit in (5.35) actually indicates a significant difference between optical links and passive networks. For example, as we just mentioned, a 1-dB passive attenuator has a noise figure of 1 dB, but (5.35) predicts that the best noise figure a link with 1 dB of loss can have is 5.13 dB; real links have even higher noise figures! Thus one way to view (5.35) is as yet another reminder that an analog link is not simply a collection of passive components. Further, it is important to obtain experimental verification that the combined form of these two limits presented in (5.35) is indeed true. We present these results in the next section.

5.5.3 General passive match limit

To apply the combined limit expressed in (5.35) to actual links requires addressing the facts that realizable, passive matching circuits rarely, if ever, are lossless and achieve a perfect match. Therefore in this section we derive the expression for the noise figure limit of an analog link where the only condition on the network between the modulation source and the modulation device is that it is passive; i.e. the network may contain loss and may not achieve a perfect match. We present the derivation in two stages: first we remove the lossless constraint, while maintaining the match constraint; then we remove both the lossless and matching constraints.

5.5.3.1 Lossy, passive match limit

Figure 5.14 is a schematic of an analog link with a lossy, passive matching network connecting the modulation source to the modulation device. The matching network loss is represented by the resistors R_1 and R_2. Continuing to use a magnitude match,

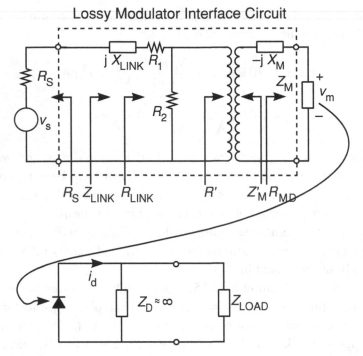

Figure 5.14 Schematic of an analog link with lossy, passive matching network between the modulation source and the modulation device. (Cox and Ackerman, 1999, Fig. 6. © 1999 International Union of Radio Science, reprinted with permission.)

we now require $X_{\text{LINK}} = 0$ and $R_{\text{LINK}} = R_{\text{S}}$, where

$$R_{\text{LINK}} = R_1 + (R_2 || R') = R_1 + \frac{R_2 R'}{R_2 + R'}. \tag{5.36}$$

As a check on the validity of (5.36), we note that when there is no matching network loss, i.e. when $R_1 = 0$ and $R_2 = \infty$, equation (5.36) reduces to the lossless magnitude match case, i.e. $R_{\text{LINK}} = R' = N_{\text{MD}}^2 R_{\text{MD}}$.

Following the development of Ackerman *et al.* (1998) we choose to express the gain of the link with lossy impedance matching, $g_{\text{i-lossy}}$, as the product of the gain with lossless matching times the loss of the matching network, g_{m}, viz.: $g_{\text{i-lossy}} = g_{\text{i}} g_{\text{m}}$. Straightforward network analysis reveals that the matching network loss can be written in terms of the resistor values:

$$g_{\text{m}} = \frac{R_{\text{LINK}} - R_1}{R_{\text{LINK}} + R_1}. \tag{5.37}$$

Alternatively, the values of the resistors representing the matching network loss can be written in terms of the matching network loss, a form that is useful when one is trying to model the measured matching network loss:

$$R_1 = \left(\frac{1 - g_m}{1 + g_m}\right) R_{\text{LINK}}; \quad R_2 = \left(\frac{4g_m}{1 - g_m^2}\right) R_{\text{LINK}}. \tag{5.38}$$

In addition to affecting the gain of the link, the addition of matching network loss also introduces additional noise sources; there are two for the network shown in Fig. 5.14. For the noise figure calculation we need to refer both these additional noise sources back to the link input. The noise source corresponding to R_1 is already at the link input whereas the noise source corresponding to R_2 only needs to be divided by the matching network loss. Making these additions produces a new expression for n_{add} (Ackerman, 1998):

$$\begin{aligned}
n_{\text{add}} = {} & \frac{1 - g_m}{1 + g_m} \frac{R_{\text{LINK}}}{R_S} kT g_{\text{i-lossy}} + \frac{1 - g_m}{1 + g_m} \frac{|Z_{\text{LINK}} + R_S|^2}{4 R_{\text{LINK}} R_S} \\
& \times \frac{[(1 - g_m)R_{\text{LINK}} + (1 + g_m)R_S]^2 + (1 + g_m)^2 X_{\text{LINK}}^2}{(R_{\text{LINK}} + R_S)^2 + X_{\text{LINK}}^2} kT g_i \\
& + \frac{R_{\text{MD}}^2 |Z_{\text{LINK}} + R_S|^2}{g_m R_{\text{LINK}} R_S |R_{\text{MD}} + Z_M'|^2} kT g_i + \left(i_{\text{rin}}^2 + i_{\text{sn}}^2\right) R_{\text{LOAD}} + kT. \tag{5.39}
\end{aligned}$$

Substituting (5.39) into (5.16) yields an expression for the noise figure of a link with lossy, passive matching (i.e. condition (5.36)):

$$\begin{aligned}
NF = {} & 10 \log \left(1 + \frac{1 - g_m}{1 + g_m} + \frac{1}{g_m}\left(\frac{1 - g_m}{1 + g_m}\right)\right. \\
& \left. + \frac{1}{g_m} + \frac{R_{\text{LOAD}}}{g_{\text{i-lossy}} kT}\left(i_{\text{rin}}^2 + i_{\text{sn}}^2\right) + \frac{1}{g_{\text{i-lossy}}}\right). \tag{5.40}
\end{aligned}$$

This expression is really not that different from the noise figure expression with lossless, passive matching, (5.34) – a fact that is more readily apparent when algebra is used to combine the first four terms in (5.40); the result is

$$NF = 10 \log \left(\frac{2}{g_m} + \frac{R_{\text{LOAD}}}{g_{\text{i-lossy}} kT}\left(i_{\text{rin}}^2 + i_{\text{sn}}^2\right) + \frac{1}{g_{\text{i-lossy}}}\right). \tag{5.41}$$

The combined limit for noise figure with lossy, passive matching at the input can be obtained from (5.41) by letting the gain be very large and assuming that the RIN and shot noise terms are negligible:

$$\lim_{\substack{i_{\text{rin}}^2 + i_{\text{sn}}^2 \to 0 \\ g_{\text{i-lossy}} \to \infty}} NF_{\substack{\text{lossy} \\ \text{match}}} = 10 \log \left(\frac{2}{g_m}\right). \tag{5.42}$$

We refer to the limit in (5.42) as the *general passive match limit*.

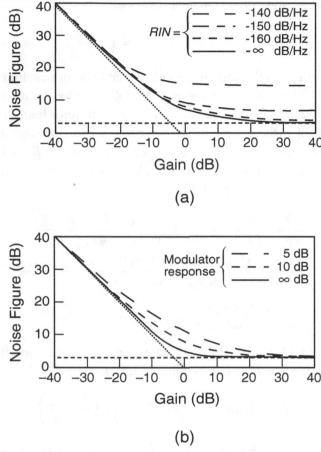

Figure 5.15 Plot of noise figure vs. intrinsic gain, (5.35), with (a) RIN as the parameter with fixed modulator response; (b) modulator response with negligible RIN.

Equation (5.42) makes clear that the only effect on the noise figure limits of having a lossy matching network is to increase the minimum noise figure from 3 dB by the amount of the matching network loss. Said differently, we can think of a link with lossy input matching as the cascade of a link with lossless matching preceded by an attenuator whose attenuation is equal to the matching network loss.

We now have a link noise figure model that is sufficiently detailed for an experimental test. From (5.35) it is clear that we need a link with high modulator response as well as negligible RIN and shot noise. To get a feel for how low the RIN needs to be, we have plotted (5.35) in Fig. 5.15(a) with RIN as the parameter and for a fixed modulator response, $N_D^2 R_{\mathrm{LOAD}}/g_i kT = 5$ dB. From this plot we see that a RIN below about -160 dB/Hz should have a negligible impact on noise figure for this experiment. For the reasons discussed in Section 5.3.3, at present only external modulation links that use a diode-pumped, solid state laser have negligible RIN. It

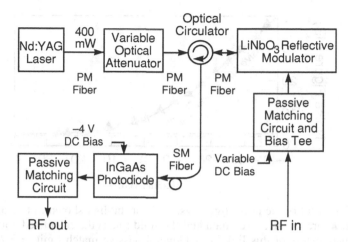

Figure 5.16 Block diagram of an externally modulated link that uses a reflective Mach–Zehnder modulator to achieve low V_π and a diode-pumped, solid state laser to achieve low RIN and high average optical power. (Cox *et al.*, 1996, Fig. 4. © 1996, IEEE, reprinted with permission.)

is also interesting to note from Fig. 5.15(a) that RIN has the same effect as matching circuit loss on the noise figure vs. gain plot.

To minimize the effects of shot noise in such links, equation (5.20) suggests that using a modulator with as low a V_π as possible is the most effective approach, since V_π appears raised to the second power in (5.20). Operating the link at as high an average optical power as possible is also important, although less so since this term only appears to the first power in (5.20). In Fig. 5.15(b) we present another plot of (5.34), this time with modulator response as the parameter and negligible RIN. From this plot we see that the effect of increasing modulator response is to sharpen the bend in the noise figure vs. gain curve around the transition between the attenuation and match limits.

A block diagram of the link assembled along these lines is shown in Fig. 5.16. A reflective form of a Mach–Zehnder modulator (Buckley and Sonderegger, 1991) is used to achieve the low V_π. Since the light passes the electrodes twice in such a device – once in the normal direction and once in the "backward" direction following reflection from a mirror – the V_π is half that of a standard Mach–Zehnder modulator with the same electrode length. The effective V_π was reduced below the actual modulator V_π by using a conjugate circuit match between the modulation source and the reflective modulator. The high power, diode-pumped, solid state laser supplied 400 mW of 1.3 μm optical power in a single-mode, polarization-maintaining fiber. An optical attenuator was used to vary the average optical power, and thereby the gain, of the link. The optical circulator was used to separate the CW and modulated optical powers.

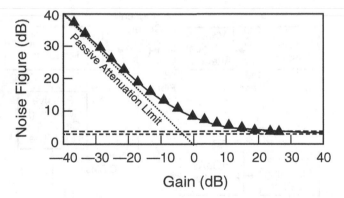

Figure 5.17 Plot of the noise figure vs. gain for the link shown in Fig. 5.16. The symbols, ▲, are the measured data and the solid line is the calculated noise figure for the parameters of this link. --- General passive match limit = 3.7 dB; ····· lossless passive match limit = 3.0 dB. (Cox *et al.*, 1996, Fig. 2. © 1996 IEEE, reprinted with permission.)

The measured noise figure of this link is plotted in Fig. 5.17 as a function of the link gain. In a separate experiment, the matching network loss was measured to be 0.7 dB, which implies that the general passive match limit for this link, (5.42), is $3 + 0.7 = 3.7$ dB. This calculated curve is also plotted in Fig. 5.17. As we can see there is excellent agreement between the measured and calculated results.

The implication of these limits for the link designer is that to reduce the link noise figure it is imperative to increase the intrinsic link gain. Further it requires substantial positive link gain to achieve a reasonably low noise figure. A lossless link, i.e. one with $G_i = 0$ dB, or even a link with a few dB of gain, generally does not have a sufficiently low link noise figure.

5.5.3.2 *Lossy, passive limit*

Up to this point we have assumed that a perfect match existed at the input between modulation source and modulation device. A common technique in low noise design is to introduce some intentional impedance mis-match to reduce the noise figure. In this section we wish to explore the changes in noise figure of an analog link as a function of the degree of input match. As we did in the preceding section, we continue to assume that the input network is lossy and passive.

Since mis-match is most commonly done in the context of a conjugate match, we also depart from the previous practice in this chapter of using a magnitude match. The resulting schematic for a link with a lossy, passive and potentially mis-matched network between modulation source and modulation device is the same as shown in Fig. 5.14, except that we now do not require the match condition, (5.36), to hold, nor do we require $X_{LINK} = 0$.

As we will see in the following, the noise figure equation for the lossy, mis-match case is so complex that it obscures the mechanism by which mis-match can reduce the noise figure. Therefore we first present the noise figure with mis-match assuming a lossless network. It has been shown by Ackerman *et al.* (1998) that permitting an input mis-match produces a modified form of the noise figure expression with perfect match (5.34):

$$NF = 10 \log \left[1 + \frac{R_{MD}^2 |Z_{LINK} + R_S|^2}{R_{LINK} R_S |R_{MD} + Z'_{MD}|^2} + \frac{(i_{rin}^2 + i_{sn}^2) R_{LOAD}}{g_i kT} + \frac{1}{g_i} \right]. \quad (5.43)$$

The significant aspect of (5.43) is the fact that the constant of 2 that appeared in (5.34) has been replaced by 1 plus a term that depends on the degree of mis-match. Under a perfect match, i.e. $R_{LINK} = R_S$, the second term in (5.43) equals 1 and consequently (5.43) reduces to (5.34). Thus to achieve a noise figure less than 3 dB requires designing the interface network such that the second term in (5.43) is less than 1. The design challenge is complicated by the fact that the mis-match also decreases the intrinsic gain.

The fact that we are using conjugate matching that is potentially mis-matched to the source implies that we need to generalize the link gain expression as well. Straightforward, albeit tedious, circuit analysis of the case where $X_{LINK} \neq 0$ and $R_{LINK} \neq R_S$ leads to the following expression for the intrinsic link gain:

$$g_i = 4 g_m \frac{R_{LINK} R_S}{|Z_{LINK} + R_S|^2} s_{md}^2 r_d^2, \quad (5.44)$$

where we have used the general subscript "md" to denote either of the modulation devices. This expression for the intrinsic gain can be used in (5.43) when $g_m = 1$, i.e. when the matching circuit is lossless.

From (5.44) it is clear that introducing the mis-match also decreases the intrinsic gain. Thus for the mis-match approach to be successful at achieving a noise figure less than 3 dB requires sufficient "excess" intrinsic gain that some can be traded away to achieve the desired noise figure reduction.

If we now permit matching circuit loss and an impedance mis-match, Ackerman *et al.* (1998) have shown that the expression for the noise figure becomes

$$NF = 10 \log \left[1 + \frac{1 - g_m}{1 + g_m} \frac{R_{LINK}}{R_S} + \frac{1 - g_m}{1 + g_m} \frac{|Z_{LINK} + R_S|^2}{4 g_m R_{LINK} R_S} \right.$$
$$\times \frac{[(1 - g_m) R_{LINK} + (1 + g_m) R_S]^2 + (1 + g_m)^2 X_{LINK}^2}{(R_{LINK} + R_S)^2 + X_{LINK}^2}$$
$$\left. + \frac{R_M^2 |Z_{LINK} + R_S|^2}{g_m R_{LINK} R_S |R_M + Z'_M|^2} + \frac{1}{g_{i\text{-lossy}}} + \frac{(i_{rin}^2 + i_{sn}^2) R_{LOAD}}{kT g_{i\text{-lossy}}} \right]. \quad (5.45)$$

An experimental demonstration of a link in which passive, lossy impedance mis-match was used to achieve a noise figure of less than 3 dB has been reported (Ackerman *et al.*, 1998). The link was identical to the one that was used to confirm the specific shape of the noise figure vs. gain curve, Fig. 5.16, with the exception that for the present purposes the modulator interface network contained an adjustable inductor and capacitor. These components permitted the degree of input impedance match to be adjusted.

For a link impedance of $Z_{LINK} = 10.6 - j19.3$ ohms at $f = 130$ MHz, the link noise figure was measured to be 2.5 dB. Due to the difficulty in measuring Z'_M Ackerman *et al.* were unable to calculate the noise figure they would have expected for this degree of mis-match.

Appendix 5.1 Minimum noise figure of active and passive networks

In this appendix we explore in more detail the basis for the passive match limit and how an active network avoids it. We begin to answer this question by considering the passive network shown in Fig. A5.1(a). Since we have assumed a passive network, the only sources within this "passive" network are thermal noise sources. As we have not assumed any particular topology for this network, the locations of these thermal noise sources are unknown. However, it turns out that it is possible (see for example Pettai, 1984, Chapter 6) to represent the combined effects of all the sources internal to a network by two sources external to the network, one at the input and one at the output. These external sources can be a pair of either voltage or current sources; if we choose to represent the network in terms of impedance parameters, [z], then the external sources are voltage sources, as shown in Fig. A5.1(b).

Using the impedance parameters of the network, we can express the voltages and currents at each of the terminal pairs for the network shown in Fig. A5.1(b) as

$$v_1 = z_{11}i_1 + z_{12}i_2 + v_{1n}, \tag{A5.1a}$$

$$v_2 = z_{21}i_1 + z_{22}i_2 + v_{2n}. \tag{A5.1b}$$

The magnitude of each of the external sources can be found by open circuiting both of the ports and measuring the resulting voltages, viz:

$$v_{1n} = v_1|_{i_1=i_2=0} \quad \text{and} \quad v_{2n} = v_2|_{i_1=i_2=0}. \tag{A5.2}$$

Further, since we have assumed that we are limiting consideration to a passive network, the external sources must be thermal sources, consequently

$$v_n = v_{1n} = v_{2n} = \sqrt{4kTR\Delta f}, \tag{A5.3}$$

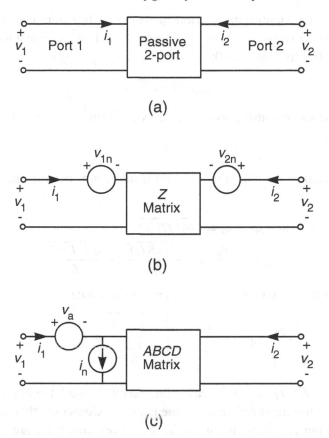

Figure A5.1 (a) Passive two-port; (b) z-parameter representation of passive two-port with all internal thermal noise sources represented by equivalent noise voltage sources at each port; (c) $ABCD$-parameter representation of passive two-port where the thermal noise voltage source at port 2 has been transformed to an equivalent thermal noise current source at port 1.

where we have also assumed that the network has the same impedance at both ports. This is always possible in a passive network because if the port impedances are not equal, then they can be made equal by the addition of an ideal transformer whose turns ratio is selected to make the port impedances equal.

By Thevenin's theorem there must be a resistor of value R across the input of each port; i.e. the real parts of z_{11} and $z_{22} = R$. It is possible via further network manipulations to have both the sources at the same port and no sources at the other. For our purposes here we transform the output voltage source to an equivalent current source, i_n, across the input port as shown in Fig. A5.1(c). The network equations for this new network are commonly written in terms of the $ABCD$ parameters:

$$v_1 = Av_2 - Bi_2 + v_a, \tag{A5.4a}$$

$$i_1 = Cv_2 - Di_2 + i_n. \tag{A5.4b}$$

We now need to establish the relationship between the external sources in (A5.1) and (A5.4). To do so we set parts (a), of (A5.1) and (A5.4) equal to one another under the constraints that $v_2 = 0$; $i_2 = 0$:

$$z_{11}i_1 + v_n = v_a \qquad (A5.5a)$$

Applying these same constraints to (A5.4b) we obtain

$$i_1 = i_n. \qquad (A5.5b)$$

Using superposition we obtain the desired relationships between the two pairs of sources:

$$v_n|_{i_1=0} = v_a = \sqrt{4kTR\Delta f}, \qquad (A5.6a)$$

$$i_n|_{v_n=0} = \frac{v_a}{z_{11}} = \frac{\sqrt{4kTR\Delta f}}{R} = \frac{\sqrt{4kT\Delta f}}{R}. \qquad (A5.6b)$$

The ratio of these two sources forms a noise resistance

$$R_{\text{noise}} = \frac{v_n}{i_n} = \sqrt{\frac{4kTR\Delta f}{\frac{4kT\Delta f}{R}}} = R. \qquad (A5.7)$$

We see from (A5.7) that in the case of a passive network the noise resistance is simply equal to the input resistance of the network. Consequently once we choose the network input resistance to match the source resistance, we are constrained to have the same resistance value match the noise resistance.

Conversely, with an active network, the input and noise resistances are no longer constrained to be equal. For example, consider the midband noise model for a bipolar transistor. It can be shown (Motchenbacher and Fitchen, 1973) that the expressions analogous to (A5.6) for the equivalent voltage and noise sources of this type of transistor are

$$v_n = \sqrt{4kT r_{bb'} \Delta f + 2q I_C r_e^2 \Delta f}, \qquad (A5.8a)$$

$$i_n = \sqrt{2q I_B \Delta f}. \qquad (A5.8b)$$

We can see from (A5.8a) that v_n contains a shot noise term, in addition to a thermal noise term. In contrast to (A5.6b), we see that in the case of a transistor, the noise current, (A5.8b), is shot noise. Thus the noise resistance of a transistor is simply the ratio part (a) to part (b) of (A5.8):

$$R_{\text{noise}} = \frac{v_n}{i_n} = \frac{\sqrt{4kT r_{bb'} \Delta f + 2q I_C r_e^2 \Delta f}}{2q I_B \Delta f} = \sqrt{\frac{2kT r_{bb'}}{q I_B} + \frac{r_\pi^2}{\beta}}, \qquad (A5.9)$$

where we have used the facts that $I_C = \beta I_B$ and $r_\pi = \beta r_e$ in writing the second form of (A5.9).

On the other hand the input resistance of a bipolar transistor can be shown to be

$$\text{Re}(z_{11}) = r_{bb'} + r_\pi, \tag{A5.10}$$

which is clearly not equal to (A5.9).

Thus the active network introduces a degree of freedom that the passive network lacks. In particular, with an active network it is possible, through judicious design, to have the input and noise resistances be different values. For example, one could choose the input resistance to match the source resistance, but arrange the noise resistance to be a different value. Such an arrangement would allow the noise figure to be less than 3 dB. It may seem counterintuitive that an active device the transistor – that has more noise sources than a passive circuit can have a lower noise figure, but this is indeed the case.

References

Ackerman, E. I. 1998. Personal communication.

Ackerman, E. I., Cox, C. H., Betts, G. A., Roussell, H. V., Ray, K. and O'Donnell, F. J. 1998. Input impedance conditions for minimizing the noise figure of an analog optical link, *IEEE Trans. Microwave Theory Tech.*, **46**, 2025–31.

Armstrong, J. A. and Smith, A. W. 1965. Intensity fluctuations in GaAs laser emission, *Phys. Rev.*, **140**, A155–A164.

Bridges, W. B. 2000. Personal communication.

Buckley, R. H. and Sonderegger, J. F. 1991. A novel single-fiber antenna remoting link using a reflective external modulator, *Proc. Photonic Systems for Antenna Applications Conf. (PSAA) 2*, Monterey.

Coldren, L. A. and Corzine, S. W. 1995. *Diode Lasers and Photonic Integrated Circuits*, New York: John Wiley & Sons, Figure 5.18.

Cox, C. H. 1992. Gain and noise figure in analogue fibre-optic links, *IEE Proc. J*, **139**, 238–42.

Cox, C. H., III and Ackerman, E. I. 1999. Limits on the performance of analog optical links. In *Review of Radio Science 1996–1999*, ed. W. Ross Stone, Oxford: Oxford University Press, Chapter 10.

Cox, C. H., III, Ackerman, E. I. and Betts, G. E. 1996. Relationship between gain and noise figure of an optical analog link, *Proc. 1996 IEEE MTT-S International Microwave Symposium*, paper TH3D-2, San Francisco, CA.

IRE 1960. IRE standards on methods of measuring noise in linear two ports, *Proc. IRE*, **48**, 60–8. Note: the IRE was one of the two professional organizations that merged to form the present IEEE.

Johnson, J. B. 1928. Thermal agitation of electricity in conductors, *Phys. Rev.*, **32**, 97–109.

Lax, M. 1960. Fluctuations from the nonequilibrium steady state, *Rev. Mod. Phys.*, **32**, 25–64.

McCumber, D. E. 1966. Intensity fluctuations in the output of CW laser oscillations, *Phys. Rev.*, **141**, 306–22.

Motchenbacher, C. D. and Fitchen, F. C. 1973. *Low-Noise Electronic Design*, New York: John Wiley & Sons.

Nyquist, H. 1928. Thermal agitation of electric charge in conductors, *Phys. Rev.*, **32**, 110–13.

Onnegren, J. and Alping, A. 1995. Reactive matching of microwave fiber-optic links, *Proc. MIOP-95*, Sindelfingen, Germany, pp. 458–62.

Petermann, K. 1988. *Laser Diode Modulation and Noise*, Dordrecht, The Netherlands: Kluwer Academic Publishers, Chapter 7.

Pettai, R. 1984. *Noise in Receiving Systems*, New York: John Wiley & Sons.

Prince, J. L. 1998. Personal communication.

Robinson, F. N. H. 1974. *Noise and Fluctuations in Electronic Devices and Circuits*, London: Oxford University Press, Chapter 4.

Schottky, W. 1918. Uber spontane Stromschwankungen in verschiedenen Elektrizitatsleitern, *Ann. Phys.*, **57**, 541–67.

Van Valkenburg, M. E. 1964. *Network Analysis*, 2nd edition, Englewood Cliffs, NJ: Prentice-Hall, Inc.

Yamada, M. 1986. Theory of mode competition noise in semiconductor injection lasers, *IEEE J. Quantum Electron.*, **22**, 1052–9.

Yamamoto, Y. 1983. AM and FM quantum noise in semiconductor lasers – Part I: Theoretical analysis, *IEEE J. Quantum Electron*, **19**, 34–46.

Yamamoto, Y., Saito, S. and Mukai, T. 1983. AM and FM quantum noise in semiconductor lasers – Part II: Comparison of theoretical and experimental results for AlGaAs lasers, *IEEE J. Quantum Electron.*, **19**, 47–58.

6

Distortion in links

6.1 Introduction

In Chapter 5 we explored one type of extraneous signals in links – noise – that because of its random nature is characterized by its statistical properties. In this chapter we investigate the other type of extraneous signals in links – distortion. Unlike noise however, distortion signals are deterministic. A further distinction between noise and distortion is the fact that while noise is always present, independent of whether there are any signals present, distortion is only present when at least one signal is present. We continue in this chapter a theme of this book by using one model to describe the distortion of both direct and external modulation, although the detailed nature of the distortion will depend on the particular modulation method that is used.

The discussion that begins this chapter is general in that the results apply to all devices with some non-linearity. The general results include the frequencies at which distortion products occur, the measures of distortion and the conversions among them. We then apply these tools to the characterization of the distortion produced by the modulation and photodetection devices that we have been studying throughout this book. For some applications the distortion levels are unacceptably high. This has led to the development of a variety of linearization techniques. The chapter concludes with an examination of two linearization techniques.

An optical link as defined in this book consists of linear passive electrical and optical components as well as modulation and photodetection devices. Since by definition the linear components cannot contribute to the link distortion, this leaves the modulation and photodetection devices as the only candidates for distortion generation. Further, it turns out that the distortion of the modulation device often dominates over the photodetection device distortion. Therefore the modulation device distortion – and methods to reduce it – will receive the emphasis in this chapter, although photodetection device distortion is also discussed.

6.2 Distortion models and measures

6.2.1 Power series distortion model

Up to this point we have only examined the linear behavior of the link components, which has been sufficient for the discussions of gain, bandwidth and noise figure. But since we now wish to consider distortion, we need to look at the non-linear characterization of the link components.

To provide some motivation for the analytical framework that is used for characterizing distortion, it may be helpful to refine our notions of what is and is not distortion (Bridges, 2001). For example, suppose we pass a square wave – as shown in Fig. 6.1(a) – through a low pass filter resulting in the waveform identified as the filtered output in this same part of the figure. The output does not resemble the input and so one might be tempted to say that the filtered output is a distorted version of the input. However, let us write down an expression for the process depicted in Fig. 6.1(a):

$$Y_{\text{out}}(s) = H(s)Y_{\text{in}}(s), \tag{6.1}$$

where $Y_{\text{in}}(s)$ and $Y_{\text{out}}(s)$ are the Laplace transform representations of the filter input and output voltages or currents, and $H(s)$ is the filter function. $H(s)$ is a complex function of s that is *independent* of the signals Y_{in} and Y_{out}. Consequently $H(s)$ can change the amplitudes and phases of the spectrum of $Y_{\text{in}}(s)$ – which means that the corresponding time function $y_{\text{out}}(t)$ can look very different from that of the input, $y_{\text{in}}(t)$ – but $H(s)$ cannot *create outputs at new frequencies*. This point is immediately apparent when we view the frequency domain representation of these same two waveforms, as is shown in Fig. 6.1(b).

This is the distinction we are driving for: we say that there is no "distortion" if no new frequencies are created. In fact, we could recover the correct shape of the waveform by passing $Y_{\text{out}}(s)$ (or $y_{\text{out}}(t)$) through an "equalizing" filter with transfer function $[H(s)]^{-1}$. This is equivalent to saying that the output will look exactly like the input if H is a constant – albeit perhaps a complex one in the mathematical sense – independent of frequency.

The reality for most electronic devices is that $Y_{\text{out}}(s)$ depends on the amplitude of $Y_{\text{in}}(s)$ in some fashion other than the simple linear proportionality expressed in (6.1). For example, consider the large-signal transfer function of a Mach–Zehnder modulator that is biased at the quadrature point, modulated with a sine wave (see (3.23)) and detected by a distortionless photodiode of responsivity r_{D}:

$$i_{\text{D}} = \frac{r_{\text{D}} T_{\text{FF}} P_{\text{I}}}{2} \left(1 + \cos\left(\frac{\pi}{2} + \frac{\pi v_{\text{m}} \sin(\omega t)}{V_{\pi}}\right)\right). \tag{6.2}$$

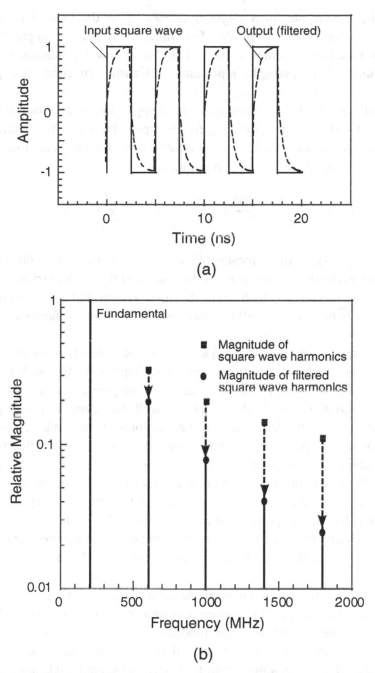

Figure 6.1(a) Time domain plot of a square wave and output after filtering through a low pass filter. (b) Frequency domain representation of these same two waveforms.

It can be shown that for a *single* frequency, ω, at the input, $i_D(t)$ will contain spectral components at ω, 3ω, 5ω, 7ω,... with relative amplitudes given by Bessel functions of order 1, 3, 5, 7,... Since new frequencies are present at the output, this example represents a transfer function that produces distortion.

For small-signals, the most common way to approximate the non-linear transfer function is via the Taylor series (see for example Thomas, 1968, Section 18.3), which expresses a function, $h(x)$, as an infinite sum of derivatives of the function with respect to x evaluated at a given point, a:

$$h(x) = \sum_{k=0}^{\infty} \frac{(x-a)^k}{k!} \left(\frac{d^k h}{dx^k}\right)_{x=a} = 1 + \sum_{k=1}^{\infty} (x-a)^k a_k, \qquad (6.3)$$

where $a_k = \frac{1}{k!}(\frac{d^k h}{dx^k})_{x=a}$. In a typical application of (6.3) to a modulation or photodetection device, the device bias point is the point a in the Taylor series. The Taylor series gives us another way to distinguish between linear and non-linear devices: a linear device is one in which all the derivatives of its transfer function other than the $k=1$ derivative are zero.

The alert reader will note that the above discussion has been in terms of voltages or currents, whereas in all the previous chapters in this book the discussion has been in terms of power in general and the power gain in particular. We will continue to use power in this chapter as well. However, the power gain only makes sense for the fundamental. For the multiples of the fundamental, there is no corresponding input power at that frequency, so the "gain" at that frequency would be infinite. Consequently, in discussions of distortion it is almost universal to discuss the power of the fundamental and its various harmonics as a function of the input power at the fundamental. But it is important to keep in mind that to actually calculate the power at the various multiples of the fundamental, one would need to first calculate the power series in terms of either voltage or current, then square the coefficients of each frequency term to obtain the power at that frequency.

We need to clarify a sometimes subtle point about distortion. A device could have a perfectly linear transfer function over some range of inputs and still produce distortion – if the magnitude of the modulation exceeds the linear range of the device. One common form of such distortion is clipping. One would experience clipping distortion, for example, with a directly modulated diode laser if the bias point and modulation were chosen such that some portion of the modulation drove the laser current below threshold. We will return to clipping distortion in the CATV portion of Section 6.2.2.

6.2.2 Measures of distortion

Equation (6.3) makes clear that the magnitude of the distortion signals, of whatever order, are not fixed quantities but depend on the magnitude of the input signal. It will be useful in the following to define the modulation depth or index, m. Recall from Chapter 2 that typically the expressions for the incremental modulated optical power – (2.14), (2.24) and (2.32) – contained a term that was the ratio of the modulating voltage or current to a maximum modulation range voltage or current. Common practice is to designate this ratio as the *modulation depth*:

$$m \equiv \frac{\text{modulating signal } (v \text{ or } i)}{\text{maximum modulation range } (V \text{ or } I)}. \tag{6.4}$$

It is clear from (6.4) that $0 < m < 1$. The modulation depth is particularly useful in the study of distortion because it turns out that most of the distortion measures can be expressed in terms of it.

Following the development of Betts *et al.* (1986), we define the transfer function of the modulation device as $h(f_{\mathrm{M}}(t))$, where $f_{\mathrm{M}}(t)$ is either the voltage or current modulation of the device. Clearly to evaluate (6.3) completely we will need the explicit functional form of h. However, by keeping the function unspecified at this point, we can develop several attributes of distortion that apply to *all* modulation and photodetection devices.

We begin by assuming that the modulation is a pure sinusoid; viz. $f_{\mathrm{M}}(t) = F_{\mathrm{B}} + f_{\mathrm{m}} \sin(\omega t)$. Substituting this expression into the second form of (6.3) and evaluating (6.3) up to $n = 3$ we obtain

$$
\begin{aligned}
h(f_{\mathrm{M}}) &\approx 1 + h_1 f_{\mathrm{m}} \sin(\omega t) + h_2 (f_{\mathrm{m}} \sin(\omega t))^2 + h_3 (f_{\mathrm{m}} \sin(\omega t))^3 \\
&= 1 + h_1 f_{\mathrm{m}} \sin(\omega t) + \frac{h_2 f_{\mathrm{m}}^2}{2}(1 - \cos(2\omega t)) \\
&\quad + \frac{h_3 f_{\mathrm{m}}^3}{4}(3 \sin(\omega t) - \sin(3\omega t)),
\end{aligned}
\tag{6.5}
$$

where we have used standard trigonometric identities to transform the first version of (6.5) into the second.

Examining the coefficient of the fundamental in (6.5), we conclude that a particularly useful form would be to let $f_{\mathrm{m}} = m/h_1$, since this would normalize the modulation to the maximum modulation range of the modulation device. Making this substitution and collecting all the coefficients of the same frequency yields the entries in the columns headed "harmonic" in Table 6.1.

Although in principle the single sinusoid stimulus can tell us everything we need to know about the power series representation of a device, a far more common RF practice is to use two equal-amplitude sinusoids for characterizing distortion:

Table 6.1 *Taylor series coefficients for both harmonic (single frequency) and intermodulation (two frequency) modulation*

	Harmonic (single frequency)		Intermodulation (two frequency)	
	Magnitude	Frequency	Magnitude	Frequency
Dc	$1 + \frac{h_2}{2}\left(\frac{m}{h_1}\right)^2$	0	$1 + h_2\left(\frac{m}{h_1}\right)^2$	0
Fundamental	$m + \frac{3h_3}{4}\left(\frac{m}{h_1}\right)^3$	ω	$m + \frac{9h_3}{4}\left(\frac{m}{h_1}\right)^3$	ω_1, ω_2
Second-order	$-\frac{h_2}{2}\left(\frac{m}{h_1}\right)^2$	2ω	$h_2\left(\frac{m}{h_1}\right)^2$	$\omega_1 \pm \omega_2$
Third-order	$-\frac{h_3}{4}\left(\frac{m}{h_1}\right)^3$	3ω	$\mp\frac{3h_3}{4}\left(\frac{m}{h_1}\right)^3$	$2\omega_1 \pm \omega_2,$ $2\omega_2 \pm \omega_1$

$f_M(t) = F_B + f_m[\sin(\omega_1 t) + \sin(\omega_2 t)]$. Making this substitution into (6.3), evaluating again up to $n = 3$ and applying standard trigonometric identities along with $f_m = m/h_1$ we obtain

$$
h(f_M) = 1 + m[\sin(\omega_1 t) + \sin(\omega_1 t)] + h_2\left(\frac{m}{h_1}\right)^2\left[\frac{1}{2}(1 - \cos(2\omega_1 t))\right.
$$

$$
+ \frac{1}{2}(1 - \cos(2\omega_2 t)) + \cos(\omega_2 - \omega_1) + \cos(\omega_2 + \omega_1)\bigg]
$$

$$
+ h_3\left(\frac{m}{h_1}\right)^3\left[\frac{9}{4}(\sin(\omega_1) + \sin(\omega_2)) - \frac{1}{4}(\sin(3\omega_1) + \sin(3\omega_2))\right.
$$

$$
- \frac{3}{4}[\sin(2\omega_1 + \omega_2) + \sin(2\omega_2 + \omega_1)
$$

$$
- \sin(2\omega_1 - \omega_2) - \sin(2\omega_2 - \omega_1)]\bigg]. \tag{6.6}
$$

The harmonic terms for each frequency in (6.6) are of course identical to the single harmonic term in (6.5). However, when two frequencies are present simultaneously, additional distortion terms are generated. These are referred to as the *intermodulation* terms. For second-order they occur at the sum and difference of the two modulation frequencies; for the third-order they occur at the sum and difference of twice one frequency and the other frequency. The coefficients for both these orders of intermodulation terms are also listed in Table 6.1.

It is important to note that although both harmonic and intermodulation distortions arise from the same underlying mechanism, they are not equal in amplitude. Comparing the appropriate cells in Table 6.1, we see that the second-order intermodulation terms are 3 dB larger than the second harmonics; the third-order intermodulation terms are 4.77 dB larger than the third harmonics.

Careful examination of both (6.5) and (6.6) reveals that the integer multiples in harmonic distortion and the sum of the integer multipliers in intermodulation distortion are equal to the exponents of the power series term that created them. Therefore all the distortion terms that came from the quadratic power series term have coefficients that sum to two: harmonics at twice either input frequency; intermodulation signals at one times each input frequency for a sum of two. Similarly the signals that came from the cubic term all equal three: three times either input frequency for the harmonic distortion and two times one frequency and one times the other – for a total of three – for the intermodulation signals. This observation gives rise to a shorthand way of referring collectively to the distortion signals produced by a given term in the power series. *Second-order distortion* refers to the harmonic and intermodulation distortion signals that arise from the quadratic term in (6.3), *third-order distortion* refers to all the distortion signals that arise from the cubic term in (6.3) and so forth. When non-linear terms greater than $n = 3$ are included, a two-tone signal will produce distortion terms at frequencies $k\omega_1 \pm l\omega_2$ for all possible integer values of k,l. Note that such distortion terms fall near ω_1 and ω_2 when k,l differ by unity.

From Table 6.1 we see that one side effect of distortion is to produce additional signals at dc and the fundamental. One way to appreciate the basis for this effect is to realize that since (6.6) is valid for all combinations of ω_1 and ω_2, it is valid when $\omega_1 = \omega_2$, i.e. when there is only a single frequency present. In this case (6.6) predicts that the second-order distortion will generate a term at $\omega_1 - \omega_1 = 0$; i.e. at dc. Under these same conditions third-order distortion will generate a term at $2\omega_1 - \omega_1 = \omega_1$; i.e. at the fundamental. If the non-linearity is small, these contributions can be negligible. However, if they are not then these additional terms can cause problems. For example, a strong second-order non-linearity can upset a bias control loop by introducing a signal dependent component to the dc bias.

From the harmonic coefficients listed in Table 6.1, we see that the magnitude of the fundamental is linearly proportional to m, at least when the fundamental contribution from the third-order distortion is negligible compared to the fundamental. Also from this table, we see that the second and third harmonic distortions are proportional to m^2 and m^3 respectively. Further, the second-order distortion is proportional to h_2, i.e. the coefficient of the quadratic term in the power series, and similarly for the third-order.

An implication of these observations is that increasing m increases the signal-to-noise ratio, since the signal increases with m, whereas the noise is independent of m. However, increasing m increases the distortion at a faster rate than it is increasing the signal, since the distortion increases as a power of m, but the signal only increases linearly with m. As long as the distortion products are low enough not to be of concern – typically this means that they are below the noise floor – then

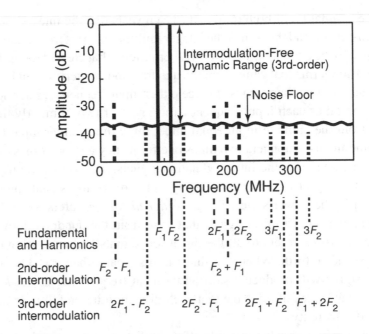

Figure 6.2 Illustrative spectrum plot of the output signals from an RF component with non-linear transfer function. The input signals are assumed to be two pure sine waves at frequencies of 90 and 110 MHz.

increasing m is advantageous. For sufficiently large m, the distortion products will rise above the noise floor, which usually sets the upper limit on the usable range of m. Consequently, the design of a link for maximum signal-to-noise ratio (with the distortion terms at or below the noise floor) involves not only the selection of the appropriate device – and perhaps linearization technique – but also matching the linear range of the device to the range of expected modulation signals, as represented by m.

Figure 6.2 is an illustrative spectrum plot of the outputs from a RF device with non-linear distortion. For the purposes of this plot the input modulation frequencies were assumed to be 90 and 110 MHz. We have arbitrarily set the amplitude of the fundamentals to 0 dB. The exact amplitudes of the harmonic and intermodulation signals depend on the non-linearity of the RF component and would be reflected in a_k coefficients in (6.3). In making Fig. 6.2 we have assumed that there is a moderate amount of distortion, so these signals are easily seen.

On inspection of Fig. 6.2 we note that all the second-order distortion signals occur at frequencies that are widely separated from the input frequencies. In the limit as the separation between the input frequencies goes to zero, the nearest second-order distortion signals are at zero and twice the input frequency. Thus over an octave bandwidth around the input frequency there are no second-order distortion signals.

Most of the third-order distortion signals are also located at frequencies that are widely separated from the input frequencies. However, there are two third-order intermodulation signals that are separated from the input frequencies by an amount equal to the spacing between the input frequencies. Thus, unlike the case with second-order distortion, there is no useable bandwidth that is free of third-order distortion.

This observation has a powerful implication. For narrowband systems, i.e. ones whose bandwidth is less than an octave, the second-order distortion can be filtered out leaving only the third-order distortion with which to contend. Wide-band systems, i.e. ones with greater than an octave bandwidth, must contend with both second- and third-order distortion.

Of course in general the power series representation of a particular device includes higher-order terms than the third. However, these higher orders are often less important, for two reasons. One is that the factorial term in the denominator of (6.3) increasingly attenuates the higher orders. The other reason that higher distortion orders are less important is that for "mildly" non-linear device transfer functions that are useful for analog modulation, the higher-order derivatives are smaller as well. Consequently, for the purposes of the discussion to follow, we concentrate on distortion orders only up to the third.

A common way to examine the effects of distortion is to plot the various output signal powers in (6.6) as a function of the input signal power. Since these signals can cover a wide range of powers, it is customary to use a log scale for both axes. A representative form of such a plot is shown in Fig. 6.3.

Unlike a linear plot where the slope represents the gain, on a log-log (or dB) plot the slope represents the exponent of the term being plotted. Thus the fundamental output varies linearly, or to first-order, with the input power. Consequently the fundamental output has a slope of one in Fig. 6.3 regardless of the gain.[1] Similarly the second-order signals are proportional to the input power squared, so these terms have a slope of two on the log-log plot, etc.

The power at which the fundamental and one of the distortion curves intersect is one measure of distortion. As we will see below, practical systems are always operated below this point to avoid serious distortion, but the intersection point is a useful measure of system performance. Clearly the higher the intercept point, the lower the distortion at a given power. In general there will be a separate intercept point for the second-order, IP_2, and third-order, IP_3, distortions, as shown in Fig. 6.3. The intercept point can be designated by either the input or output power at which it occurs. But since these powers differ by the gain of the device, it is important to be clear as to which definition is being used.

---

[1] The gain is the intercept on the vertical axis at 0 dB on the horizontal axis.

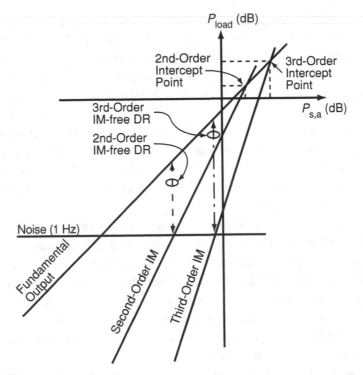

Figure 6.3 Plot of the output signal powers – fundamental, second- and third-order intermodulation – as a function of the input signal power. Note that log scales are used for both axes.

The intercept point is a measure of distortion only, i.e. it is not affected by the noise figure of the device. In most devices it is not possible to measure the intercept point directly, typically because the output begins to saturate before intersecting the input curve. In such cases the intercept point is defined as the point where a linear extrapolation of the two curves would intersect.

Another measure of distortion is the *intermodulation-free dynamic range*, IM-free DR. Like the intercept point, there is a different IM-free DR for each distortion order. However, unlike the intercept point, the IM-free DR is a function of both the distortion and the noise of the component. Thus to define the IM-free DR we need to add the noise power to Fig. 6.3. Recall that the output noise power is independent of the input signal power, and therefore a horizontal line in Fig. 6.3 represents the noise.

The minimum magnitude of the fundamental, p_{min}, that a link can convey is when the *fundamental* power is equal to the noise power. The maximum magnitude of the fundamental that a link can convey without any measurable distortion, p_{max}, is when the modulation produces *distortion* power equal to the noise power. In principle one could use any of the distortion terms discussed above in the definition of IM-free

DR. In practice, by far the most common distortion terms used are the second- and third-order intermodulation terms. Consequently the second- and third-order IM-free DR is defined as p_{max}/p_{min} or $P_{max} - P_{min}$ in dB, where p_{max} depends on the IM distortion order. The second- and third-order IM-free DRs are also plotted in Fig. 6.3.

Since the IM-free DR is dependent on noise power, which in turn is dependent on bandwidth, the IM-free DR is bandwidth dependent. As with noise, it is often convenient to quote the IM-free DR in a 1-Hz bandwidth. The user can then scale the 1-Hz number to the application bandwidth, BW, using the appropriate bandwidth scaling law. However, because of the various slopes involved, there is a different scaling law for each order of IM-free DR and all are a little more complex than the corresponding law for noise.

Consider first the bandwidth scaling law for second-order IM-free DR, IMF_2. As shown in Fig. 6.4(a), a 1-dB increase in the noise power results in a 1-dB increase in the fundamental, but only a $\frac{1}{2}$-dB increase in the second-order intermodulation power. Thus the net decrease in the IM-free DR is $\frac{1}{2}$ dB. Therefore the second-order IM-free DR scales as $(b\omega)^{\frac{1}{2}}$, or $(BW)/2$ in dB, as does the second-order harmonic distortion, HD_2. For example, to convert second-order IM-free DR from a 1-Hz bandwidth to a 1-MHz bandwidth, we first calculate the ratio of these two bandwidths: $1/10^6$, which corresponds to $BW = -60$ dB. Consequently IMF_2 (1 MHz) $= IMF_2$ (1 Hz) $- 30$ dB.

A similar line of reasoning can be used to establish the third-order, IM-free DR (IMF_3), scaling law. From Fig. 6.4(b) we see that a 1-dB increase in noise power causes only a $\frac{1}{3}$-dB increase in third-order intermodulation power for a net decrease in IM-free DR of $\frac{2}{3}$ dB. Consequently the third-order IM-free DR scales as $(b\omega)^{\frac{2}{3}}$, or $\frac{2}{3}(BW)$ in dB, as does the third-order harmonic distortion, HD_3. Continuing the example from above, to convert third-order IM-free DR from a 1-Hz bandwidth to a 1-MHz bandwidth, we can use the same bandwidth ratio of -60 dB; thus we can immediately write IMF_3 (1 MHz) $= IMF_3$ (1 Hz) $- 40$ dB.

From these two examples we conclude that the scaling law for an arbitrary distortion order, n, is $(b\omega)^{\frac{n-1}{n}}$ or $\frac{n-1}{n}(BW)$.

From Fig. 6.3 it is also clear that the two distortion measures we have discussed can be related by the simple formulae:

$$IMF_2 = \tfrac{1}{2}(IP_2 - N_{out}); \quad IMF_3 = \tfrac{2}{3}(IP_3 - N_{out}), \tag{6.7}$$

where $N_{out} = G_t + NF + 10\log(kT\Delta f/10^{-3})$ and IP_n, the output powers at the intercept points are in dBm; the IMF terms are in dB in the bandwidth Δf.

The second- and third-order IM-free distortion measures are most useful in applications where the input signal spectrum consists of a few, irregularly spaced

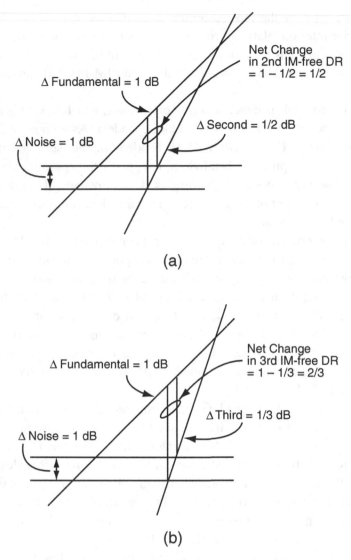

Figure 6.4 Plot of output fundamental and (a) second-order or (b) third-order distortion vs. fundamental modulation power.

signals. Typical applications include radar and communication systems. In such applications it is unlikely that a fundamental will occur at the same frequency as the intermodulation signal from another pair of signals.

There are also applications where multiple, regularly spaced carriers are frequency multiplexed onto the same optical link; cable TV (CATV) signals are one example, cellular telephone signals are another. However, recall that two of the third-order intermodulation terms are separated in frequency from the carrier by the separation of the carriers. Thus a third carrier that continues the equal spacing of

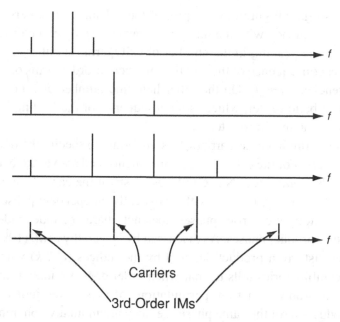

Figure 6.5 Representative spectral plot of a frequency multiplexed system with regularly spaced carriers showing how multiple carrier pairs can produce third-order intermodulation at the same frequency.

the pair will fall on top of one of the intermodulation terms from the original pair. Conversely, as shown in Fig. 6.5, the regular carrier spacing in these applications implies that there exist multiple carrier pairs that will produce intermodulation signals at the *same* frequency. In turn these multiple signals can interfere and add to levels that are *greater* than simple power addition. In such cases there is a need for a new distortion measure that makes statistical assumptions about the semi-coherent addition of the multiple intermodulation terms. The common measures that are often used are the *composite second-order*, CSO, and the *composite triple beat*, CTB.

Like the IM-free DR, the CSO/CTB distortion measures are defined as the difference in powers (in dBm) between the carrier and distortion signals. The difficulty in implementing this definition directly is that for either of these distortion measures, the regular carrier spacing means that the strongest distortion signals are located at the carrier frequencies. Therefore to measure CSO or CTB, the carrier that would otherwise mask the distortion signal is turned off.

Unlike the IM-free DR, the CSO/CTB distortion measures are a function of where they are measured within the passband of the system. The tedious task of counting up all the CSO/CTB terms for a given frequency multiplexing format and plotting them vs. frequency is ideally suited to a computer, or a motivated and meticulous freshman. Figure 6.6 is a plot of the calculated numbers of CSO and CTB terms vs. frequency with the number of carriers as a parameter (Phillips and Darcie, 1997). The peaking in the number of CSO terms at either end of the pass band is a direct consequence of the fact that the second-order terms occur at widely spaced frequencies: $f_i \pm f_j$. On the other hand the number of CTB terms peaks around the pass band center, which is a consequence of the folding back into the band of some of the third-order terms.

With multiple simultaneous carriers, it is important to specify the relative phases among the carriers. For the cable TV example mentioned above, the National Television Standards Committee (NTSC) places most[2] of the carriers on a spacing of $109.25 + 6n$ MHz, $n = 0, 1, 2, \ldots$ with statistically independent phases. However, independence among the carrier phases does not guarantee independence among all the distortion product phases. As discussed by Nazarathy *et al.* (1993), consider one particular distortion product denoted by the indices (i, j, k) with $i \neq j \neq k$. Elementary combinatorics tells us that when order does not matter, there are five other distortion products that are permutations of these three indices. These six distortion products have the same phase, i.e. they are mutually coherent and hence these terms add in amplitude making their sum six times larger than a distortion term where the order of the indices does matter.

It can be shown (Phillips and Darcie, 1997) that the theoretical CSO and CTB can be expressed in terms of the modulation depth and the power series coefficients (as summarized in Table 6.1), just as we did for harmonic and intermodulation distortion:

$$CSO = 10 \log \left(N_{\text{CSO}} h_2^2 m^2 \right), \tag{6.8a}$$

$$CTB = 10 \log \left(\frac{9}{4} N_{\text{CTB}} h_3^2 m^4 \right), \tag{6.8b}$$

where N_{CSO} and N_{CTB} are the numbers of second-order and third-order products for a specified number of carriers and at a specific frequency in the band as shown in Fig. 6.6.

To emphasize the fact that CSO/CTB are simply another way of characterizing the same underlying non-linearity we characterized with the IM-free DR, we can use the coefficients in Table 6.1 to derive dependences similar to (6.8) for the second- and third-order IM-free-DRs. Equations (6.8a, b) are in terms of RF powers; consequently we need to square the coefficients in Table 6.1, since they were derived in terms of voltage or current. Further, since intermodulation – like CSO/CTB – measures the distortion relative to the fundamental, we need the

---

[2] There are five carriers below this frequency, at 55.25, 61.25, 67.25, 77.25 and 83.25 MHz. Due to other constraints, such as the commercial FM radio band, the carriers are not regularly spaced.

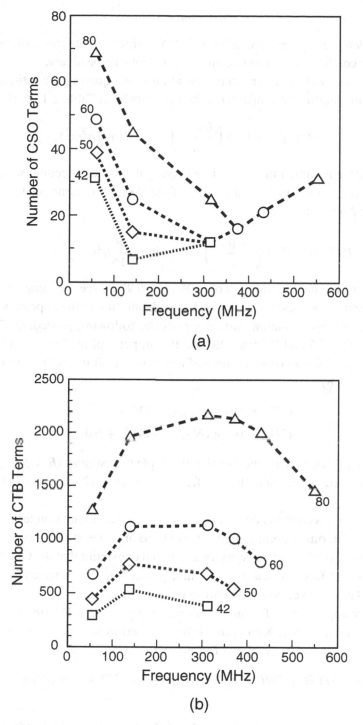

Figure 6.6 Plot of the number of (a) CSO and (b) CTB terms vs. frequency with the number of carriers as the parameter. (Data from Tables 14.1 and 14.2, respectively, of Phillips and Darcie, 1997.)

intermodulation suppressions, IMS_2 and IMS_3, which are the ratios of second- and third-order coefficients to fundamental coefficients, respectively.

Consider the second-order intermodulation first; squaring the ratio of second-order intermodulation to fundamental coefficients from Table 6.1 we obtain

$$IMS_2 \propto 10 \log \left(\frac{h_2 m^2}{m} \right)^2 = 10 \log \left(h_2^2 m^2 \right), \tag{6.9a}$$

where we have assumed that $h_1 = 1$. Applying the same procedure we can obtain the analogous expression for the ratio of third-order intermodulation power to fundamental power:

$$IMS_3 \propto 10 \log \left(\frac{3 h_3 m^3}{4m} \right)^2 = 10 \log \left(\frac{9}{16} h_3^2 m^4 \right). \tag{6.9b}$$

Comparing the analogous parts of (6.8) and (6.9) we see the same dependencies on power series coefficients and modulation depth. In fact the appearance of identical terms in these equation pairs suggests the following procedure: To convert between CSO/CTB and IM-free DR, use the property of the logarithm function to express the products in (6.8) as sums of logs and substitute (6.9) in for the second log terms in (6.8):

$$CSO = 10 \log(N_{CSO}) + IMS_2 \tag{6.10a}$$

$$CTB = 10 \log(N_{CTB}) + IMS_3 + 6 \, \text{dB}, \tag{6.10b}$$

where the 6 dB arises from the fact that the triple beat stimuli $(f_1 \pm f_2 \pm f_3)$ used in the CTB are 6 dB larger than the two frequency stimuli $(2f_1 \pm f_2)$ used to measure IM-free DR.

In fact this conversion is correct as long as the intermodulation terms are above the noise so that one is comparing the CSO/CTB and intermodulation *suppression*. However, the IM-free DR imposes an additional constraint that the CSO/CTB does not: for the IM-free DR the fundamental power must be selected such that the intermodulation power equals the noise power.

There are approximate formulae for converting between IM-free DR and CSO/CTB (Betts *et al.*, 1991; Kim *et al.*, 1989). For example, the IMF_3 can be related to the *CNR*, *CTB* and N_{CTB} by the following formula:

$$IMF_3 \, (1 \, \text{Hz}) \cong \tfrac{2}{3}(CNR + 10 \log(\Delta f)) + \tfrac{1}{3}(|CTB| + 6 + 10 \log(N_{CTB})). \tag{6.11}$$

The first term on the right hand side of (6.11) converts the bandwidth, Δf, of the *CNR* to the 1-Hz bandwidth of the IMF_3. The second term corrects for the additional power that arises from summing the multiple intermodulation terms vs. the two terms of the IMF_3. We have added an absolute value sign to the *CTB* term because while it is common to express *CTB* as a negative number, it is not universal. In any case we need a positive *CTB* to make (6.11) work. There are also two important conditions to using (6.11). One is that the same modulation depth must be used when obtaining the *CNR* and *CTB* values. The other is that (6.11) was derived before linearization became popular; consequently it is only valid for third-order distortion.

Representative values for a typical CATV application are *CNR* = 55 dB, *CTB* = 65 dB, Δf = 4 MHz and N_{CTB} = 3025 (a typical number for a 100 channel system), which results in an IMF_3 = 116 dB Hz$^{2/3}$. As a second check on (6.11) we have used an RF amplifier data sheet (WJ, 1998), which lists both IP_3 and *CTB*. Using (6.7) we can convert the given IP_3 of 41 dBm into IMF_3 (1 Hz) = 133 dB. Using (6.11) we can convert the given *CTB* of –61 dBc into IMF_3 (1 Hz) = 133 dB (the data sheet lists the *CTB* for 110 channels, which we assumed was close enough to 100 that we used N_{CTB} = 3025).

6.3 Distortion of common electro-optic devices

6.3.1 Diode laser

Throughout this section we assume that optical reflections of the laser's own light back into the laser cavity are negligible; the degrading effects of such optical reflections on both distortion and noise will be treated in Chapter 7.

For the first part of this section we also assume that the modulation frequency is much less than the laser relaxation frequency. Under this assumption it turns out that the distortion can be accurately predicted from the shape of the laser's dc *P* vs. *I* curve. In other words, in this frequency range the laser distortion is independent of the laser's dynamics.

Recall from Section 2.2.1 that the slope efficiency was defined as the derivative of the laser *P* vs. *I* curve in (2.5), which we repeat here for convenience:

$$s_\ell(i_{\text{L}} = I_{\text{L}}) \equiv \left. \frac{\partial p_\ell}{\partial i_\ell} \right|_{i_\ell = I_{\text{L}}}. \tag{6.12}$$

Subsequently in Chapter 3 we showed that the intrinsic gain of a directly modulated link is proportional to slope efficiency, i.e. $g_t \propto s_\ell$. It follows that the "gain" of the *n*th-order term is proportional to the *n*th-order derivatives of the laser *P* vs. *I* curve:

$$g_{t,n}(i_{\text{L}} = I_{\text{L}}) \propto \left. \frac{\partial^n p_\ell}{\partial i_\ell^n} \right|_{i_\ell = I_{\text{L}}}. \tag{6.13}$$

Further, the laser current is proportional to the available power from the modulation source, i.e. $i_\ell \propto p_{\text{s,a}}$. Consequently for a diode laser we can investigate the distortion by expressing the power series, (6.3), for the laser optical power as a function of the laser current and the derivatives of the laser P vs. I curve (Petermann, 1988):

$$p_\ell = \frac{\partial p_\ell}{\partial i_\ell} i_\ell + \frac{1}{2} \frac{\partial^2 p_\ell}{\partial i_\ell^2} i_\ell^2 + \frac{1}{6} \frac{\partial^3 p_\ell}{\partial i_\ell^3} i_\ell^3 + \cdots, \tag{6.14}$$

where we have continued to assume that we are operating the laser about a bias current I_L, although this is not shown explicitly in (6.14).

In principle one needs a different power series for every bias point on the P vs. I curve to characterize completely the distortion of a laser diode. In practice, as long as the combination of bias and modulation currents does not result in any portion of the modulation signal driving the laser below threshold, the relevant power series is only a weak function of bias point. Distortion arising from this constraint is referred to as *small-signal distortion*.

There are modulation formats, most commonly the subcarrier multiplexing format used for the distribution of CATV signals, where the composite modulation signal does occasionally drive the laser below threshold. Since the P vs. I curve slope changes radically around threshold, we would require a significantly different power series to predict the resulting distortion. This general class of distortion is referred to as *large-signal distortion*; in the particular case of a diode laser it is more commonly referred to as clipping. In the following we limit the discussion to small-signal distortion.

Even without an analytic expression for the P vs. I curve, we can make some qualitative statements about the bias dependence of a laser's distortion from (6.14). Consider the representative P vs. I curve shown in Fig. 6.7(a). Equation (6.14) indicates that the fundamental gain – which is proportional to the slope of the P vs. I curve – is relatively constant both above and below threshold, but with higher gain above threshold; see Fig. 6.7(b).[3] From (6.14) the second-order distortion is proportional to the rate of change of the P vs. I curve slope. i.e. to the curvature of the P vs. I curve. This suggests that the second-order distortion has a maximum around lasing threshold as shown in Fig. 6.7(c), then decreases to a relatively constant value above threshold. Since the second-order distortion goes through an inflection point at the maximum, the third derivative vanishes at this point. This fact implies that the third-order distortion will be minimum around threshold, as shown in Fig. 6.7(d).

Measurements of distortion on actual diode lasers confirm the general features we have just described. The data shown in Fig. 6.8 are a plot of the second- and third-order distortion vs. average optical power. Near threshold the second-order

[3] It is important to keep in mind that a diode laser's optical power is approximately a linear function of the *current*. Since the source is typically a voltage source in series with a 50-Ω resistor, this constraint will be met as long as the diode laser resistance is small compared to 50 Ω.

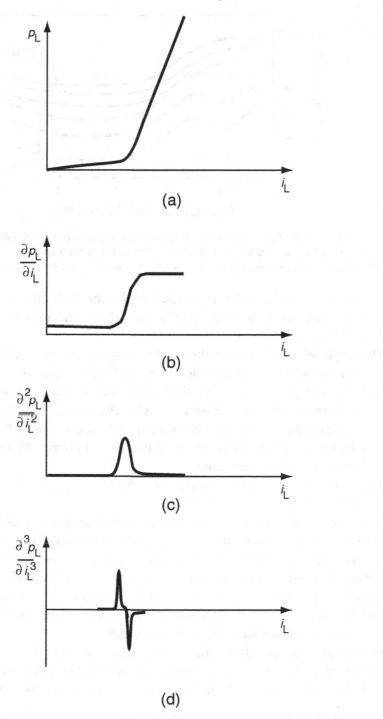

Figure 6.7 Sketch of typical laser diode *P* vs. *I* curve, (a) and its first three derivatives, (b), (c) and (d) respectively.

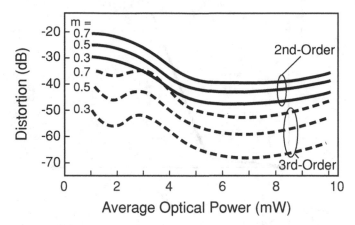

Figure 6.8 Plot of experimentally measured second- and third-order distortion vs. average optical power with modulation index as the parameter (Petermann, 1986, Fig. 4.19. © Kluwer Academic Publishers, reprinted with permission.)

distortion is greater than the third-order. Above threshold, the second-order decreases and remains relatively constant while there appears to be no regular pattern to the third-order distortion.

The implication of these results for designing a directly modulated link with minimum distortion is that there is no "perfect" bias where one or more of the distortion orders goes to zero[4] while the fundamental is not zero. Consequently, from a distortion perspective, the bias point should be chosen sufficiently above threshold to maximize the gain and to avoid the peaking of the second-order distortion around threshold, but not so far above threshold that the gain begins to decrease and the distortion begins to increase again.

The above results are general in the sense that they apply to all of the diode laser structures described in Section 2.2.1.

It is tempting to assume that one could establish an estimate of a laser diode's distortion by simply taking the derivatives of an experimentally measured P vs. I curve. However, the low levels of distortion and the inherently "noisy" process of taking derivatives combine to make it presently impractical to predict laser distortion based on a measured P vs. I curve. Consequently one can obtain better results by simply applying a modulation signal to the diode laser and measuring the harmonics and intermodulation distortion directly.

In principle one could use the analytic expression for the P vs. I curve that was derived in Appendix 2.1 to predict the dc and low frequency distortion of a diode laser. This is generally not done because at frequencies of interest to most

---

[4] It is true that around threshold the third-order distortion goes to zero, but this bias point is not of practical importance since the gain is low or rapidly changing with current around this bias point.

link designers, the distortion effects of the laser dynamics cannot be neglected. Therefore the more useful approach is to extend the ac rate equations derived in Appendix 4.1 to include the effects of distortion. The derivation of these expressions from the rate equations is straightforward, but tedious. Thus we defer the derivation to Appendix 6.1 and present only the results below, which are similar to those originally presented by Darcie *et al.* (1985). As was the case when we derived the diode laser frequency response at the fundamental, the modulation index, m, and the ratio of the modulation to relaxation frequencies are key parameters. The explicit dependencies on these parameters of second harmonic, third harmonic and third intermodulation distortion, normalized to the fundamental response, are:

$$\frac{HD_2}{F} \cong 10\log\left\{ m\frac{\left(\frac{f_1}{f_r}\right)^2}{g(2f_1)} \right\}, \tag{6.15a}$$

$$\frac{HD_3}{F} \cong 10\log\left\{ \frac{3m^2}{2}\frac{\left(\frac{f_1}{f_r}\right)^4 - \frac{1}{2}\left(\frac{f_1}{f_r}\right)^2}{g(2f_1)g(3f_1)} \right\}, \tag{6.15b}$$

$$\frac{IMF_3}{F} \cong 10\log\left\{ \frac{m^2}{2}\frac{\left(\frac{f_1}{f_1}\right)^4 - \frac{1}{2}\left(\frac{f_1}{f_1}\right)^2}{g(f_1)g(2f_1)} \right\}, \tag{6.15c}$$

where the gain of the laser medium as a function of frequency is given by

$$\frac{1}{g(f)} = \left\{ \left(\left(\frac{f}{f_r}\right)^2 - 1\right)^2 + \left(\frac{2\pi\varepsilon f}{g_0}\right)^2 \right\}^{-\frac{1}{2}}, \tag{6.16}$$

in which g_0 is the uncompressed gain and ε is the gain compression factor due to photon density, N_P. These parameters can be combined into an empirical expression for the material gain that includes compression:

$$g(N_P) = \frac{g_0}{1 + \varepsilon N_P}. \tag{6.17}$$

Figure 6.9 shows a plot of (6.15a, b, c) for the case where $f_r = 5.3$ GHz. The value of ε/g_0 was selected for the best match between the experimental fundamental frequency response and (6.17). Notice that all distortion measures are increasing from their low frequency values and that they all peak at a modulation frequency

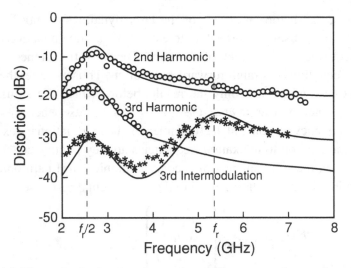

Figure 6.9 Plots of calculated and experimentally measured second- and third-order harmonic distortion as well as third-order intermodulation distortion vs. modulation frequency for a laser relaxation frequency of 5.3 GHz (Darcie, T. E., Tucker, R. S. and Sullivan, G. J. 1985. Intermodulation and harmonic distortion in InGaAsp lasers, *Electron. Lett.*, **21**, 665–6. Reprinted with permission.)

that is half the relaxation frequency. The harmonic distortions then decrease monotonically above $f_r/2$. However, the intermodulation distortion first decreases, then increases to a higher maximum at the relaxation frequency. This is another reason why the IM-free DR is the primary measure of distortion.

Also plotted in this figure are the measured data of Darcie *et al.*, which agree quite well with the theoretically predicted curves. Similar measurements have been made on other diode laser structures (gain guided and index guided), different laser cavities (Fabry–Perot and distributed feedback) and lasers fabricated via different growth techniques. All yield distortion curves that depend primarily on the parameters given above.

It is important to keep in mind the fact that (6.15a, b, and c) are valid only at frequencies approaching the relaxation frequency. For example, setting $f_1 = 0$ in any of these equations would predict that the distortion goes to zero at low frequencies, which is in conflict with the finite low frequency distortion we examined above. A typical laser for CATV has its relaxation frequency a factor of 5–10 times the maximum modulation frequency.

6.3.2 Mach–Zehnder modulator

As we saw in Chapter 4, the modulation frequency response of a Mach–Zehnder modulator does not contain an inherent device resonance like the diode laser.

Therefore the Mach–Zehnder transfer function is – at least ideally – independent of frequency.[5] Consequently the distortion, which is dependent on the shape of the transfer function, should be independent of frequency. Betts *et al.* (1990) have shown that to within experimental error, the magnitude of the third-order intermodulation signal for a Mach–Zehnder modulator is independent of frequency out to 20 GHz.[6]

The even-order distortion is a strong function of bias, the odd-order distortion significantly less so, as was discussed by Bulmer and Burns (1983) in one of the earliest papers on the use of Mach–Zehnder modulators for analog applications. However, unlike the diode laser, we have an explicit analytic form for the Mach–Zehnder transfer function. We begin our development of the dependence of distortion on bias by recalling the expression for the Mach–Zehnder transfer function, (2.12), which is repeated below for convenience:

$$p_{M,O} = \frac{T_{FF} P_1}{2} \left(1 + \cos \left(\frac{\pi v_M}{V_\pi} \right) \right). \tag{6.18}$$

As we did with the diode laser, we take the first, second and third derivatives of (6.18) to obtain the bias dependences of the gain, second- and third-order distortions, respectively; the results are

$$\frac{\partial p_{M,O}}{\partial V_M} = -\frac{T_{FF} P_1}{2} \frac{\pi}{V_\pi} \sin \left(\frac{\pi V_M}{V_\pi} \right), \tag{6.19a}$$

$$\frac{\partial^2 p_{M,O}}{\partial V_M^2} = -\frac{T_{FF} P_1}{2} \left(\frac{\pi}{V_\pi} \right)^2 \cos \left(\frac{\pi V_M}{V_\pi} \right), \tag{6.19b}$$

$$\frac{\partial^3 p_{M,O}}{\partial V_M^3} = -\frac{T_{FF} P_1}{2} \left(\frac{\pi}{V_\pi} \right)^3 \sin \left(\frac{\pi V_M}{V_\pi} \right). \tag{6.19c}$$

Figure 6.10 shows plots of all parts of (6.19). As was observed in Section 2.2.2.1, the maximum gain for the Mach–Zehnder modulator occurs at the bias points $V_M = kV_\pi/2$, where k is an integer. At these same bias points, the second-order distortion is zero. In fact, since the Mach–Zehnder transfer function possesses odd symmetry about these bias points – i.e. $p_{M,O}(v_M - V_M) = -p_{M,O}(-(v_M - V_M))$, *all* even-order distortions are zero at these bias points.

The third-order distortion also has bias points where it is zero, i.e. $V_M = kV_\pi$. Further, since the Mach–Zehnder transfer function has even symmetry about these bias points, i.e. $p_{M,O}(v_M - V_M) = p_{M,O}(-(v_M - V_M))$ *all* odd-order distortions are zero at these bias points. Unfortunately this last statement implies that the

---

[5] For example the skin effect loss of the modulator electrodes, which is proportional to $\sqrt{f}$, would introduce a frequency dependence.
[6] The IM-free DR, however, is not independent of frequency because the decrease in modulator response at higher frequencies increases the link noise figure.

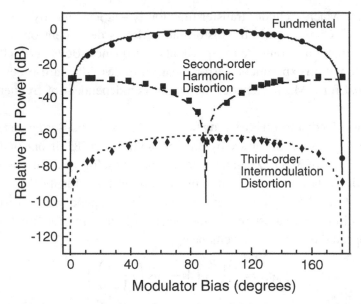

Figure 6.10 Lines are plots of all parts of (6.19), fundamental, second- and third-order distortion, respectively, vs. Mach–Zehnder bias angle. Data points are experimentally measured values of gain and distortion (Roussell, 2001). (Reprinted with permission from the author.)

fundamental gain – which is the lowest, odd-order term – is also zero at these bias points. Consequently these bias points are of no use for operation with low third-order distortion.

Thus with a standard Mach–Zehnder modulator it is possible to select a single bias point to eliminate the second- but not third-order distortion. This fact *per se* is of limited utility, since as was pointed out in Section 6.2.2 there is no bandwidth for which the second-order distortion is the only distortion term. However, as will be discussed in Section 6.4 below, this feature is quite useful when combined with a third-order linearization scheme to produce a broadband link with low second- and third-order distortion.

Also shown in Fig. 6.10 are measurements of the gain, second- and third-order distortion, which are in good agreement with the analytically predicted results.

A final note on distortion in Mach–Zehnder modulators: since the transfer function is periodic, this type of modulator does not suffer from clipping distortion like the diode laser did. However, to analyze the distortion under large-signal conditions, we would need to re-derive the distortion, beginning with the expression for large-signal gain, which for a Mach–Zehnder modulator was derived in Section 3.4.

6.3.3 Directional coupler modulator

The directional coupler is similar to the Mach–Zehnder in the sense that in both cases the distortion is a function of bias but not of frequency. However, as we shall see, there is an important difference in the dependence of distortion on bias between these two modulators.

In Section 2.2.2.2 the transfer function for the directional coupler was given by (2.19):

$$p_{D,O} = T_{FF} P_I \frac{\sin^2 \left[\frac{\pi}{2} \sqrt{1 + 3 \left(\frac{v_M}{V_S} \right)^2} \right]}{1 + 3 \left(\frac{v_M}{V_S} \right)^2}. \tag{6.20}$$

To calculate the bias point dependence of the gain, second- and third-order distortion, we once again take the first three derivatives of (6.20). However, the more complex functional form of (6.20) – as compared to (6.18) – results in considerably more complex expressions for the derivatives. Situations such as these are ideally suited to the symbolic math computer programs, such as Mathematica. Using this program, Prince (1998) has shown the first three derivatives of (6.20) to be expressible as

$$\frac{\partial p_{D,O}}{\partial V_M} = T_{FF} P_I \frac{3 V_S^2 V_M}{2 \left(V_S^2 + 3 V_M^2 \right)^2} \left[-2 + 2 \cos \left(\pi \sqrt{1 + 3 \left(\frac{V_M}{V_S} \right)^2} \right) \right.$$

$$\left. + \pi \sqrt{1 + 3 \left(\frac{V_M}{V_S} \right)^2} \sin \left(\pi \sqrt{1 + 3 \left(\frac{V_M}{V_S} \right)^2} \right) \right], \tag{6.21a}$$

$$\frac{\partial^2 p_{D,O}}{\partial V_M^2} = T_{FF} P_I \frac{3}{2 \left(V_S^2 + 3 V_M^2 \right)^3} \left[V_S^2 \left(-2 V_S^2 + 18 V_M^2 \right) \right.$$

$$+ \left(2 V_S^4 + (3\pi^2 - 18) V_S^2 V_M^2 + 9\pi^2 V_M^4 \right) \cos \left(\pi \sqrt{1 + 3 \left(\frac{V_M}{V_S} \right)^2} \right)$$

$$\left. + \pi V_S^2 \left(V_S^2 - 12 V_M^2 \right) \sqrt{1 + 3 \left(\frac{V_M}{V_S} \right)^2} \sin \left(\pi \sqrt{1 + 3 \left(\frac{V_M}{V_S} \right)^2} \right) \right], \tag{6.21b}$$

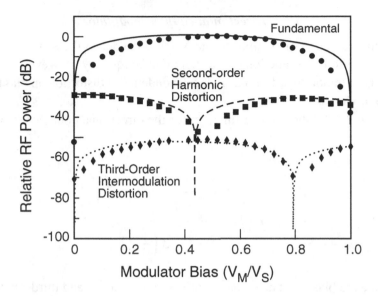

Figure 6.11 Lines are plots of all parts of (6.21), fundamental, second- and third-order distortion, respectively, vs. directional coupler bias voltage. Data points are experimentally measured values of gain and distortion (Roussell, 1995). (Reprinted with permission from the author.)

$$
\frac{\partial^3 p_{\mathrm{D,O}}}{\partial V_{\mathrm{M}}^3} = T_{\mathrm{FF}} P_{\mathrm{I}} \frac{27 V_{\mathrm{M}}}{2\left(V_{\mathrm{S}}^2 + 3 V_{\mathrm{M}}^2\right)^4} \left[-8\left(V_{\mathrm{S}}^4 - 3 V_{\mathrm{S}}^2 V_{\mathrm{M}}^2\right) + \left[(8 - \pi^2) V_{\mathrm{S}}^4 \right.\right.
$$

$$
\left. + (3\pi^2 - 24) V_{\mathrm{S}}^2 V_{\mathrm{M}}^2 + 18\pi^2 V_{\mathrm{M}}^4\right] \cos\left(\pi \sqrt{1 + 3\left(\frac{V_{\mathrm{M}}}{V_{\mathrm{S}}}\right)^2}\right)
$$

$$
+ \pi \sqrt{1 + 3\left(\frac{V_{\mathrm{M}}}{V_{\mathrm{S}}}\right)^2} \left[5 V_{\mathrm{S}}^4 + (\pi^2 - 18) V_{\mathrm{S}}^2 V_{\mathrm{M}}^2 + 3\pi^2 V_{\mathrm{M}}^4\right]
$$

$$
\left. \times \sin\left(\pi \sqrt{1 + 3\left(\frac{V_{\mathrm{M}}}{V_{\mathrm{S}}}\right)^2}\right) \right]. \tag{6.21c}
$$

Figure 6.11 is a plot of all parts of (6.21) vs. directional coupler bias. In general the curves in Fig. 6.11 appear similar to the corresponding plots in Fig. 6.10. The gain is maximum and the second-order distortion is zero at a common bias point of $V_{\mathrm{M}} = 0.438$, which is only slightly less than the 0.5 value for the Mach–Zehnder. However, the directional coupler transfer function is not periodic in bias voltage, so this bias condition does not repeat as was the case with the Mach–Zehnder. Also, since this bias point is only an inflection point and not a point about which odd symmetry exists – as it was with the Mach–Zehnder, the higher, even-order distortions are generally not zero at this bias point.

Careful examination of Fig. 6.11 reveals that there is a bias point, $V_M = 0.7954$, where the third-order distortion goes to zero while the fundamental – although down from its maximum value – is not zero. This occurs because unlike the Mach–Zehnder, the directional coupler modulator transfer function does not possess even symmetry at the bias where the third derivative goes to zero. Therefore a standard directional coupler modulator can have a large third-order, IM-free DR.[7] This feature would be useful in narrowband applications, i.e. those with less than an octave bandwidth, where the second-order distortion can be filtered out. The standard directional coupler is less useful in broadband applications because there is no *single* bias point where both the second- and third-order distortions go to zero.

The data points in Fig. 6.11 (Roussell, 1995) show the measured values of gain and distortion and are in reasonably good agreement with theory.

6.3.4 Electro-absorption modulator

Recall from Section 2.2.2.3, (2.27), that the transfer function for the electro-absorption (EA) modulator can be expressed as

$$p_{A,O} = T_{FF} P_I e^{-\Gamma L \Delta a(v_M)}. \tag{6.22}$$

In this form it would appear quite straightforward to investigate the distortion of an EA modulator by taking successive derivatives of (6.22) as we have done for the Mach–Zehnder and directional coupler modulators. The difficulty in attempting to do this stems from the fact (see Appendix 2.2) that there is an Airy function (differential form) in the expression for the absorption coefficient as a function of voltage in the exponential of (6.22). Although it is possible to express the Airy function derivatives in closed form, we do not gain much design insight from doing so. Consequently we revert to the power series approach as was suggested for the diode laser, (6.14), to study the EA distortion as a function of bias.

Welstand *et al.* (1999) have measured the transfer function of their EA modulator vs. bias and fitted it empirically with a seventh-order power series. They then used the resulting power series to calculate the second- and third-order distortion

---

[7] Since the third-order distortion goes to zero, it might be expected that the IM-free DR would be infinite at this bias point. In fact it turns out that there can be distortion power at the third-order frequency even though the third-order coefficient is zero. Briefly the reason for this effect is that higher odd-order distortion can produce distortion signals at the "third-order" frequency. Consequently, eliminating the third-order term eliminates the dominant component of the distortion signal at the third-order frequency, but it does not eliminate this signal completely unless the coefficients of the appropriate higher-order distortion terms are also zero. This point will be discussed more completely in Section 6.4.2.

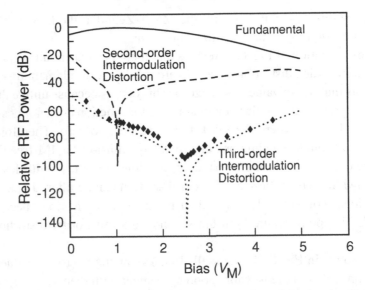

Figure 6.12 Lines are calculated fundamental, second- and third-order distortion, respectively, vs. bias; data points are experimentally measured third-order IM-free dynamic range (after Welstand *et al.*, 1999, Fig. 5).

as a function of EA modulator bias. The second- and third-order IM-free DRs that resulted from this process are plotted in Fig. 6.12. Like the directional coupler modulator, the EA modulator has bias points where either the second- or the third-order distortion is minimum while the fundamental is not at a minimum. Also plotted on Fig. 6.12 are the experimentally measured data for the third-order IM-free DR, which are in excellent agreement with the calculated results. The rationale as to why the distortion of an EA modulator has this particular bias dependence depends on the details of the modulator design and consequently is beyond the scope of this book.

6.3.5 Photodiode

A photodiode is basically a linear device in converting optical power to an electrical current. Consequently any photodiode distortion arises not from the basic transfer function but from second- or higher-order effects. Thus it is not productive to predict photodiode distortion by taking derivatives of its theoretically ideal transfer function. However, photodiode distortion has been measured (see for example Ozeki and Hara, 1976; Esman and Williams, 1990; Williams *et al.*, 1993) and modeled (see for example Dentan and de Cremoux, 1990; Williams *et al.*, 1996).

There are, of course, myriad internal photodiode parameters that affect the level of distortion it produces. Knowledge of how the photodiode distortion depends on

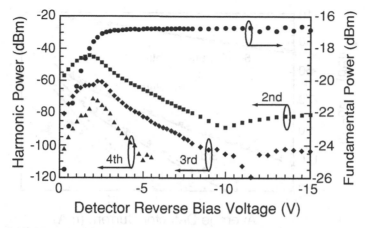

Figure 6.13 Measured fundamental (5 GHz) and harmonic RF powers for a PIN photodiode as a function of the reverse bias across the photodiode with an average photocurrent of 1 mA (Williams *et al.*, 1996, Fig. 5. © 1996 IEEE, reprinted with permission.)

these parameters is of little use to the link designer since they are not typically part of the link design process. For example, the link design process does not typically include varying the active layer doping level or the buffer layer thickness to achieve the desired performance.

However, there are two principal external photodiode parameters – which are available to the link designer – that affect the photodiode distortion. They are the reverse bias applied to, and the average optical power incident on, the photodiode.

Figure 6.13 is a plot of the measured fundamental RF power from a PIN photodiode along with the harmonic distortion powers for orders two through four as a function of the reverse bias across the photodiode (Williams *et al.*, 1996). The fundamental frequency was 5 GHz and the average photocurrent, 1 mA, was chosen to be just below the start of the range where this would also contribute to the non-linear response.

We see from the figure that as the bias is increased from zero, initially the fundamental and distortion powers increase. This initial increase with bias can be attributed to the decrease in the junction capacitance as the reverse bias increases. The smaller capacitance shunts less of the fundamental thereby allowing more to reach the photodiode load. However, a bias is soon reached where the fundamental power no longer increases. Consequently if one were merely attempting to determine a bias voltage to maximize the fundamental output power, one could choose any bias more negative than this value.

But Fig. 6.13 also makes clear that the harmonic powers continue to decrease over a significant bias range beyond the bias voltage where the fundamental is constant.

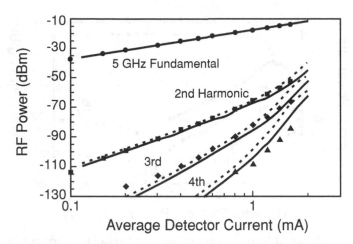

Figure 6.14 Measured and simulated fundamental (5 GHz) and harmonic RF powers for a p-i-n photodiode as a function of average photocurrent for a reverse bias of 5 V (Williams *et al.*, 1996, Fig. 7. © 1996 IEEE, reprinted with permission.)

Eventually a bias voltage is reached beyond which all the powers – harmonics as well as fundamental – are essentially independent of further increases in reverse bias. These data make clear that the reverse bias for minimum distortion is significantly higher than the bias needed to achieve maximum fundamental power from the photodiode.

To investigate the distortion dependence on average photodiode current, we fix the bias at a nominal value of −5 V and vary the average optical power incident on the photodiode. Figure 6.14 is a plot of measured data and simulation results of the fundamental (again at 5 GHz) along with harmonic distortion powers for orders two through four vs. the average photodiode current (Williams *et al.*, 1996).

For this particular photodiode, the fundamental output power did not show any saturation effects over the range of measured photodiode currents. However, as was the case with reverse bias, the harmonic powers are a more sensitive indicator of the onset of a deviation from linearity in the photodiode transfer function. Above about 1 mA we see evidence of a change in the photodiode transfer function since all the harmonic distortion powers begin to increase faster with average photocurrent than they did below this value.

Since both direct and external modulation links are often operated at high average optical power, the above results are potentially a problem for both types of links. At low modulation frequencies, the limitations of high average optical power can be largely avoided by spreading the optical power over a larger area detector. Both the detectors used by Williams *et al.* in the measurements reported above were less than 10 μm in diameter. By using a larger diameter photodiode, e.g. 50 to 300 μm,

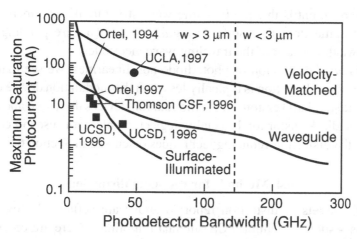

Figure 6.15 Plots of calculated maximum saturation current vs. 3-dB frequency for surface illuminated, traveling wave and velocity-matched photodiodes (Lin *et al.* © 1997 IEEE, reprinted with permission). Data points are from results reported in the literature: • velocity-matched distributed PD; ■ waveguide PD; ▲ surface-illuminated PD (Prince, 1998).

one can operate the photodiode with 30 mA of current yet still achieve modulation-device-limited distortion (Ackerman *et al.*, 1998). The basis for this technique is that power density, rather than total power *per se*, is the controlling parameter for distortion. As pointed out by Yu (1998), the maximum power density as a function of position within the detector depends on the Joule heating at that position, which in turn depends on the detector geometry. Thus in practice many factors enter into the determination of the maximum optical power for a detector.

As the modulation frequency one is attempting to detect increases, the area of a surface-illuminated photodiode needs to decrease, for the reasons discussed in Section 2.3. This is clearly at odds with the distortion-driven desire to increase the photodiode area so as to keep the power density low. Considerable effort has been devoted to the problem of increasing the maximum photocurrent of high frequency photodiodes while maintaining the photodiode linearity (see for example Giboney *et al.*, 1997; Lin *et al.*, 1997; Jasmin *et al.*, 1997; Williams *et al.*, 1996). Lin *et al.* (1997) have calculated the maximum saturation photocurrent vs. frequency for three classes of photodiode: surface illuminated, waveguide and velocity matched. Their results are plotted in Fig. 6.15 along with the experimentally reported saturation currents for a number of photodetectors (Prince, 1998). These results make clear the rapid decrease in maximum saturation current with increasing frequency for surface illuminated photodiodes.

The traveling wave photodiode structure significantly reduces the decrease in maximum saturation current as the frequency increases. One of the primary reasons for this improvement is that the traveling wave structure permits increases in the absorption volume along the optical waveguide, without a corresponding reduction in the bandwidth because of the traveling wave electrodes.

From this brief discussion of photodiode non-linearities, we see that for most photodiodes, their distortion is typically less than the modulation device distortion, at least for the standard versions of the modulation devices discussed above. Consequently, photodiode distortion is usually neglected unless it is used in conjunction with one of the distortion-reducing techniques discussed in Section 6.4.

6.4 Methods for reducing distortion

Although the levels of distortion reported above are sufficiently low for many applications – such as cellular/PCS antenna remoting – there are other common applications – such as radar and CATV – where lower distortion levels are required. To meet the needs of these applications a variety of linearization techniques, i.e. techniques for reducing the distortion, have been developed.

There are at least three criteria by which linearization techniques can be categorized. One is the domain, primarily electrical or primarily optical, in which the reduction is achieved. Another criterion is the bandwidth of the linearization technique: broadband that reduces both second- and third-order vs. narrowband, which reduces only third-order, often at the expense of increasing second-order. A third criterion is the distortion orders to which the linearization technique applies: a particular order such as the third, or all orders.

In the presentation below, we have elected to categorize the linearization techniques by the first criterion. For each technique we also point out where it lies on the other two axes as well.

Since the dominant source of distortion in a link is generally caused by the modulation device, the emphasis in developing linearization techniques to date has largely been on the modulation device. As a first thought, one might pursue the following approach to linearization: select a bias point, determine the transfer function in the vicinity of this bias point and then design a linearization scheme to reduce the total distortion to acceptable levels. In practice this turns out to be an overly hard way to approach the problem. By examining the problem a little more carefully, we find a significant simplification.

As we saw in Section 6.3, there is generally a bias point where either the second- *or* third-order distortion is zero. Thus the linearization approach usually taken is to operate at one of these bias points and then design a distortion reduction method to reduce the other distortion order. The location of the second-order distortion minimum is typically near the center of the device bias range, whereas the third-order distortion minimum – if it exists at all – is often nearer one of the extremes in the bias range. Since a centrally located bias point permits a higher modulation index, one usually chooses to operate the modulation device at a bias where the second-order distortion is minimum and then designs the linearization to reduce the third-order distortion. Most of the linearization techniques to be presented below follow this general approach.

The primarily electronic linearization methods take the modulation transfer function as given and attempt to design an electronic circuit to compensate for the overall non-linearity. The primarily optical methods take the approach of attempting to design a modulation device or combination of devices with a more linear transfer function. We now explore representative examples of each of these approaches in more detail.

6.4.1 Primarily electronic methods

The three general classes of linearization techniques where the distortion reduction is achieved primarily in the electronic domain are shown in block diagram form in Fig. 6.16. These methods have been applied to both direct and external modulation devices.

A block diagram of pre-distortion (see for example Bertelsmeier *et al.*, 1984; Wilson *et al.*, 1998), the most common approach to linearization, is shown in Fig. 6.16(a). If the non-linear transfer function that is limiting the IM-free DR is well known and stable, a circuit with the inverse of the non-linearity can be inserted before the modulation device. The cascade of the two non-linearities then has a more linear transfer function than the original single non-linearity. The distortion reduction results can be impressive. For example, Wilson *et al.* report 22.6 dB reduction in the CTB for their pre-distortion circuit with adjustable amounts of third- and fifth-order correction.

However, pre-distortion is an "open loop" approach, so any changes in the offending non-linearity will not trigger a correction in the compensating non-linearity. The results can range from less increase in IM-free DR to degradation in IM-free DR! On the positive side, pre-distortion is simple to implement and is not as limited in the frequency range over which it can be implemented, as is the case with the feedback approach to be presented next.

Although it may seem obvious, an often overlooked aspect of pre-distortion is the requirement to have the bandwidth between the pre-distorter and the modulation device be at least the order of the distortion one is trying to cancel. For example, a third-order pre-distorter needs at least three times the fundamental bandwidth, a fifth-order pre-distorter needs five times, etc. Applying this criterion to the CATV application where the fundamental bandwidth is typically 500 MHz, implies that

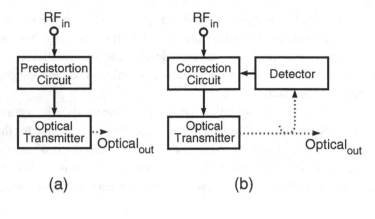

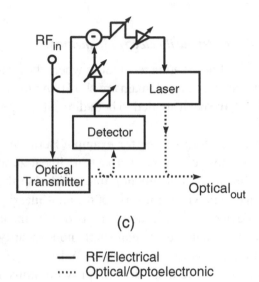

— RF/Electrical
····· Optical/Optoelectronic

Figure 6.16 Block diagrams of the three general categories of electronic linearization: (a) pre-distortion, (b) feedback and (c) feedforward.

third-order linearization requires at least 1.5 GHz of bandwidth and fifth-order 2.5 GHz.

The feedback approach (see for example Straus, 1978) to linearization, as shown in Fig. 6.16(b), can compensate for changes in the transfer function non-linearity. An overview of its operation is as follows. A portion of the optically modulated signal is detected and compared with the original electrical modulating signal. If the photodetection distortions are negligible,[8] as they typically are, then the feedback will drive the difference between these two signals to zero – i.e. the optical

---

[8] If the photodetector non-linearities are not negligible, then the feedback loop will still linearize the signal; however, the linearized signal will be at the photodetector output, rather than at the modulator output.

modulated signal will be an identical replica of the electrical modulating signal. Hence any distortion imposed by the modulating device will have been eliminated. Changes in the modulator distortion function only change the operating point of the feedback loop – but a linearized modulated signal is maintained.

For the feedback loop to drive the dc error signal to zero, we have made a common assumption about the feedback loop: that it contains a single, ideal integrator. In such feedback loops the average value of the error signal must be zero; if it were not, then the integrator would integrate up the finite error until the integrator reached its limiting value, which is generally set by the power supply.

Despite its advantages, feedback has its own limitations. Primary among them is the need to limit the phase delay around the feedback loop so that the loop is stable. Compounding this problem is the lack at present of feedback amplifiers with sufficient gain above about 400 MHz. Thus to utilize feedback at high frequencies will require a substantial development effort covering feedback amplifier design and monolithic integration of this amplifier and the modulation device.

The third general category of linearization is feedforward (see for example Frankart *et al.*, 1983; deRidder and Korotky, 1990) and is shown in block diagram form in Fig. 6.16(c). This is probably the most complex to implement because it usually requires two modulation devices. This approach relies on the fact that lower amplitude modulation signals generate significantly lower levels of distortion products. Thus the distortion introduced by modulating the first modulation device is added back, but out of phase, by the second modulation device. However, because of the lower amplitude of the distortion signal used to modulate the second device, the distortion by this device is much smaller.

One drawback of feedforward is the fact that the distortion cancellation does not occur until the two signals are combined on the photodetector as power; optical coherence is not required. This can be a problem in links that convey high frequency modulation over long lengths of fiber at wavelengths where fiber dispersion is significant. The different phase shifts in the modulation that accrue to the various frequency components can degrade the cancellation because terms that left the modulation device perfectly out of phase do not arrive at the photodetector still out of phase.

It might appear at first that the second modulation device is required to inject only a small amount of power. However, as pointed out by Nazarathy *et al.* (1993), the power of the second modulation device needs to handle the peak of the composite modulating signal if it is to correct the distortion. This is the time domain analogy to the requirement that the bandwidth of the pre-distorter needs to include the harmonics. If the modulation signal consists of a large number, N, of signals at frequencies f_i whose phases are random with respect to one another, then we can obtain an estimate of the peak by treating the composite modulation signal as a stationary Gaussian process with power spectrum:

$$S(f) = \frac{1}{2} \sum_{i=1}^{N} \left(\frac{\pi v_m}{V_\pi}\right)^2 \delta(f - f_i),$$ (6.23)

where $\delta(f)$ is the impulse function. The standard deviation of (6.23) is given by

$$\sigma = \frac{\pi v_m}{V_\pi} \sqrt{\frac{N}{2}}.$$ (6.24)

From the elementary properties of a Gaussian distribution it is known that 99% of the time the peak modulation signal, $v_{M,P}$ will fall in the range $\pm 3\sigma$. The peak distortion, Δ_P, is simply the difference between the actual and ideal linear modulator transfer functions at Δ_P. The peak-to-peak deviation is simply $2\Delta_P$. Consequently the peak power required from the secondary modulation device is $2\Delta_P P_M$, where P_M is the average optical power from the primary modulation device.

To illustrate the preceding, consider a CATV application using a Mach–Zehnder modulator. A common set of parameters for this application is $N = 60$, $m = \pi v_m / V_\pi = 3.6\%$ and $P_M = 10$ mW. Substituting these values into (6.24) yields $\sigma = 0.2$ which in turn yields $v_{M,P} = 0.6$. To calculate Δ_P we need to evaluate the difference between the actual and ideal linear modulator transfer functions. For the case of a Mach–Zehnder modulator, which has a sinusoidal transfer function, the expression for Δ_P becomes

$$\Delta_P = \sin\left(\frac{\pi v_{M,P}}{V_\pi}\right) - \frac{\pi v_{M,P}}{V_\pi}.$$ (6.25)

Substituting $v_{M,P} = 0.6$ into (6.25) yields $\Delta_P = 3.54\%$. Finally we obtain the peak power required to compensate non-linearities from the primary modulation device from $2\Delta_P P_M$, $= 0.71$ mW. Depending on the ratio used to combine the secondary and primary modulation signals, typical values for the actual peak power from the secondary modulation device could range from 1.36 mW with an ideal 3-dB tap to 2.72 mW for an ideal 20–80% tap. Excess losses in the actual tap would increase these values by the amount of the actual tap's excess loss.

The latter two techniques have in common that they can compensate for any order of non-linearity, as opposed to pre-distortion that is used primarily to cancel a particular order of non-linearity, such as third-order.

There are also hybrid approaches, which are a combination of these methods. For example one can combine pre-distortion with feedback (Nazarathy et al., 1993) as shown in Fig. 6.17. The disadvantage of pre-distortion in failing to adapt to changes in the non-linear transfer function is overcome by having the feedback loop. The frequency limitation of feedback is overcome by having a low frequency feedback loop. This is accomplished by comparing not the input and feedback signals, as

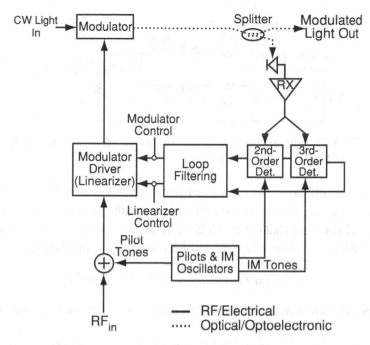

Figure 6.17 Block diagram of hybrid electronic linearizer using pre-distortion and feedback. (Nazarathy *et al.*, 1993, Fig. 19. © 1993 IEEE, reprinted with permission.)

was the case when we first discussed feedback above, but by detecting the amount of distortion and using that dc level to control the amount of pre-distortion. Such an approach is feasible if the non-linearity is frequency independent, as is the case with Mach–Zehnder modulators. Nazarathy *et al.* (1993) report 26 dB reduction in CTB for their hybrid approach.

We now explore in a little more detail the design of an electronic pre-distorter as applied to an electro-absorption modulator, following the development of Wilson *et al.* (1998). As was pointed out in Section 6.3.4, the analytical form of the EA transfer function is unwieldy. Consequently we begin by representing the EA modulator transfer function, (6.22), by a power series in the modulation voltage, v_m:

$$p_{a,o} = T_{FF} P_I \left(a_1 v_m + a_2 v_m^2 + a_3 v_m^3 + a_4 v_m^4 + \cdots \right). \qquad (6.26)$$

Following the linearization methodology presented above, we select a bias point such that $a_2 = 0$ and attempt to design a pre-distortion circuit to eliminate the third-order distortion.

The transfer function of the pre-distortion circuit can be represented by a power series in v_{in}:

$$v_o = b_1 v_{in} + b_2 v_{in}^2 + b_3 v_{in}^3 + b_4 v_{in}^4 + \cdots. \qquad (6.27)$$

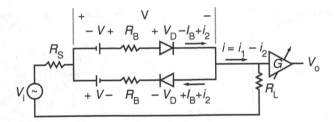

Figure 6.18 Schematic of diode pre-distortion circuit. (Wilson *et al.*, 1998, Fig. 4. © 1998 IEEE, reprinted with permission.)

Since the output voltage of the pre-distortion circuit is the electro-absorption modulation voltage, we can evaluate the performance of the cascade of these two components by substituting (6.27) into (6.26) and assuming $b_2 = 0$. For the purposes of this discussion we keep only the odd-order terms up to third-order:

$$p_{a,o} \cong T_{FF} P_I \left(a_1 b_1 v_{in} + \left(a_1 b_3 + a_3 b_1^3 \right) v_{in}^3 \right). \tag{6.28}$$

Under these constraints, the third-order distortion is eliminated when $a_1 b_3 = -a_3 b_1^3$.

The a coefficients are determined by fitting the power series to the measured EA modulator transfer function, and thus are taken as a given for the pre-distorter design. Consequently the object of the pre-distorter design is to develop a circuit that is capable of generating the desired distortion and expressing that distortion, i.e. the b-coefficients, as a function of the circuit parameters.

A common technique for generating the required pre-distortion transfer function is to use a pair of diodes connected in parallel with opposite polarity (Childs and O'Byrne, 1990), as shown in Fig. 6.18. As we show below, one advantage of using diodes to generate the pre-distortion function is the fact that a dc parameter – the bias current through the diodes – can be used to change the pre-distortion function.

For sufficiently low frequencies we can neglect the frequency-dependent components, such as the junction capacitance, of the ideal diode. In this case the relation between the voltage across an ideal diode,[9] $v_D = v_d + V_D$, and the current through the diode, $i_D = i_d + I_D$, is given by (see for example Gray and Searle, 1969):

$$i_D = I_S \left(e^{v_D / V_T} - 1 \right), \tag{6.29}$$

where I_S is the saturation current, i.e. the "leakage" current that flows under reverse bias – as is easily seen from (6.29) when $v_D \ll 0$, $i_D \cong I_S$ – and V_T is the thermal voltage given by kT/q – it is about 25 mV at room temperature.

---

[9] Real diodes follow this same equation but with a constant – called the ideality factor – which is usually denoted by m in the exponent of the exponential. Typical values of the ideality factor for silicon diodes are two to three.

To analyze the pre-distortion circuit at the modulation frequency, we can eliminate the bias voltage source, in which case we can write the expression for the diode voltage by inspection from elementary circuit theory: $v_d = v - i_d R_B$. Substituting this expression into (6.29) we obtain

$$i_d = I_S \left(e^{(v - i_d R_B)/V_T} - 1 \right), \tag{6.30}$$

which holds for each diode of the pair.

Recall that (6.27) was a power series in voltage, which means that we need the inverse of (6.30). Solving (6.30) for the voltage yields

$$\frac{v}{V_T} = \ln \left[1 + \left(\frac{i_d}{I_S} \right) \right] + \frac{I_S R_B}{V_T} \left(\frac{i_d}{I_S} \right). \tag{6.31}$$

The final power series we need is one for (6.31):

$$\frac{v}{V_T} = \left(1 + \frac{I_S R_B}{V_T} \right) \left(\frac{i_d}{I_S} \right) - \frac{1}{2} \left(\frac{i_d}{I_S} \right)^2 + \frac{1}{3} \left(\frac{i_d}{I_S} \right)^3 - \frac{1}{4} \left(\frac{i_d}{I_S} \right)^4 + \cdots. \tag{6.32}$$

In most designs of practical interest, $I_S R_B / V_T \ll 1$, which means that we can neglect this term in (6.32). Deriving an analogous expression for the other diode, combining the two and applying a process known as power series reversion, we obtain a power series for the pre-distortion circuit:

$$\frac{v_i}{V_T} \cong \left(1 + \frac{2 I_S R}{V_T} \right) \left(\frac{i_d}{2 I_S} \right) - \frac{1}{6} \left(\frac{i_d}{2 I_S} \right)^3 + \cdots, \tag{6.33}$$

where we have only given explicit expressions for the terms through third-order.

The pre-distorter output voltage – which is also the modulator voltage – is simply the output current times the load resistor, R_L. Multiplying both sides of (6.33) by the thermal voltage and matching like terms between (6.33) and (6.27) we obtain

$$b_1 = \left(1 + \frac{2 I_S R}{V_T} \right) \left(\frac{V_T R_L}{2 I_S} \right), \tag{6.34a}$$

$$b_3 = \frac{V_T}{6} \left(\frac{R_L}{2 I_S} \right)^3, \tag{6.34b}$$

where $R = R_S + R_L$.

To get a feel for the linearization that is practical with a diode pre-distortion circuit, Wilson *et al.* have measured the distortion of an EA modulator alone and in combination with a diode pre-distorter. For these measurements the EA modulator was biased for minimum second-order distortion, as measured by the CSO. At an optical modulation depth of 2.7%, the highest CSO in the CATV band that Wilson *et al.* measured was -61.3 dB at channel 2 (about 55 MHz). Since the pre-distorter

Distortion in links

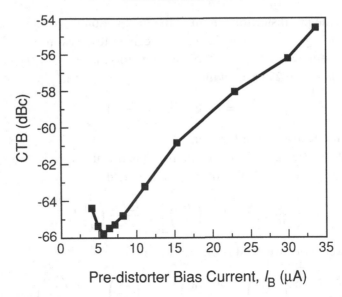

Figure 6.19 Plot of measured CTB vs. pre-distorter bias of an EA modulator with diode pre-distortion. (Wilson *et al.*, 1998, Fig. 10. © 1998 IEEE, reprinted with permission.)

is only reducing the third-order distortion, the CSO is not a function of the pre-distorter bias.

Without the pre-distortion circuit, the highest CTB in the CATV band that Wilson *et al.* measured for the EA modulator was −42.8 dB at channel 40; with pre-distortion the CTB at channel 40 was reduced to −65.4 dB – or an improvement of 24.1 dB. At the band edges the reduction in distortion was slightly less; for example at channel 2 the improvement in CTB with pre-distortion was 19.8 dB, with similar performance at channel 76.

One of the practical questions for any linearization method is: how sensitive is the performance enhancement to changes in parameters such as temperature, bias current, etc.? In the Wilson experiments the temperature of the EA modulator was stabilized by a thermo-electric cooler. Figure 6.19 is a plot of CTB for the combined diode pre-distorter and EA modulator vs. pre-distorter bias current. As can be seen from the data in the figure, there is a "best" bias current. However, there also is a range of bias currents – from about 4 to 7.5 μA – over which the CTB only degrades by 1 dB from this best value. This range of bias current is easily within the capability of a control circuit. Unfortunately no data are available on the performance of the pre-distorter vs. EA modulator bias current.

6.4.2 *Primarily optical methods*

In contrast to the primarily electronic linearization methods, the primarily optical methods have been applied almost exclusively to external modulation devices.

These methods typically involve interconnecting two modulation devices and selecting the biases such that the transfer function of the combination is more linear
than either of the individual devices. Betts *et al.* (1995) have categorized a number of the primarily optical linearization methods based on the starting modulation
device – Mach–Zehnder or directional coupler – and on whether the linearization reduces second-order only, third-order only or both. The result is shown in
Fig. 6.20.

The IM-free DR numbers listed beside some of the modulator configurations
shown in Fig. 6.20 are from the work of Bridges and Schaffner (1995), who used
computer simulations to compare the IM-free DR of various optical linearization
techniques. These data indicate that the IM-free DR can be improved by as much
as 15 dB. However, as they also discuss, the practicality of actually achieving such
improvements is critically dependent on the ability to maintain the required dc
bias and in some cases the required ratio of RF powers feeding the modulators.
The difficulty of implementing this latter requirement significantly decreases the
practicality of many of these linearization techniques.

As was discussed in the individual modulator sections, a single Mach–Zehnder
or directional coupler modulator has a bias point where the second-order distortion
is zero. To cancel simultaneously the second- and third-order distortion requires
two modulators connected either in series or parallel. The idea underlying all these
linearization methods is to arrange for the distortion produced by the second modulator to be equal in magnitude but out of phase with the distortion produced by the
first modulator, thereby canceling the distortion from the combined device. To be
useful, the fundamental signals from the two modulators must not cancel, although
invariably there is some reduction in the fundamental magnitude.

In some cases the two effective modulators may share the same physical waveguides, making it appear that there is only a single modulator. For example, the dual-
polarization Mach–Zehnder of Johnson and Roussell (1988), the dual-wavelength
Mach–Zehnder modulator of Ackerman (1999) and the multiple-electrode directional coupler modulator of Farwell *et al.* (1991) fall in this category.

Of the primarily optical linearization methods, we choose to examine in detail the
series connection of two Mach–Zehnder modulators. In the version first proposed
and demonstrated by Skeie and Johnson (1991) the second and third y-branch
couplers between the two modulators and the fourth y-branch coupler at the output
of the second modulator were replaced with coupling regions. For applications
where one needs to balance the output powers from each branch of the fourth
coupler, they later included the ability to control the coupling ratios via biases
applied to the electrodes alongside each of the coupling regions. In such designs –
see Fig. 6.21(a), there are many degrees of freedom: the bias points of the two
modulators, the coupling ratios of the variable couplers and the RF power split
between the two modulators. In a typical CATV application, the coupling ratios

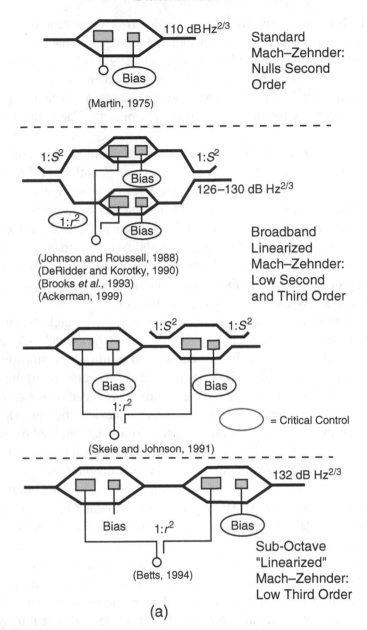

Figure 6.20 Illustrated table of primarily optical linearization methods: (a) Mach–Zehnder-based and (b) directional-coupler-based. Each of the methods is further categorized by the distortion order(s) reduced (Betts *et al.*, 1995).

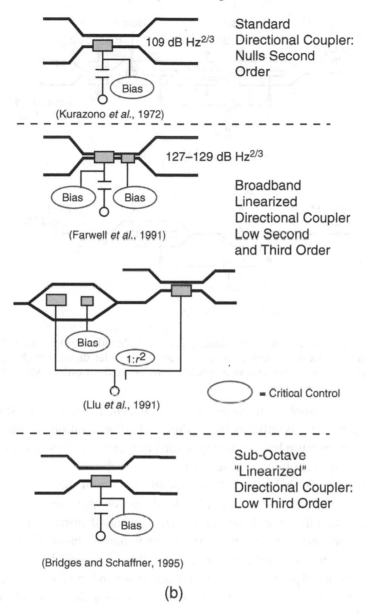

Standard Directional Coupler: Nulls Second Order

109 dB Hz$^{2/3}$

(Kurazono et al., 1972)

Broadband Linearized Directional Coupler Low Second and Third Order

127–129 dB Hz$^{2/3}$

(Farwell et al., 1991)

1:r^2

(Liu et al., 1991)

⬭ = Critical Control

Sub-Octave "Linearized" Directional Coupler: Low Third Order

(Bridges and Schaffner, 1995)

(b)

Figure 6.20 (cont.)

and the power split were chosen to optimize performance while the two modulator dc biases were controlled to linearize both the second- and third-order distortion.

As might be imagined with all these biases, this is a relatively complex modulator to analyze. Therefore we will investigate a simpler version of the dual series Mach–Zehnder modulator first proposed and demonstrated by Betts (1994). In the Betts

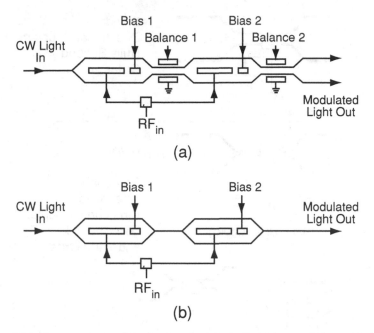

Figure 6.21 Two versions of the dual series Mach–Zehnder modulator: (a) broadband version capable of reducing second- and third-order distortion; (b) narrowband (less than an octave) version capable of reducing third-order distortion only.

version, the two variable ratio couplers are eliminated. The remaining structure, as shown in Fig. 6.21(b), is still capable of third-order linearization and consequently this modulator structure has found application in systems with less than an octave bandwidth, such as is often the case in radar.

In the development below, we show that this version of the dual series Mach–Zehnder actually increases the second-order distortion. A natural question at this point is: given that the broadband version reduces both distortion orders, why not use this more general linearization method even in narrow-band applications? The answer involves bias control complexity as well as a tradeoff between distortion reduction and noise figure – topics which both go beyond the focus of this chapter, which is on distortion only. The distortion/noise figure tradeoffs will be discussed in Chapter 7.

Before plunging into the mathematical description of this modulator, it is helpful to understand intuitively how this modulator increases the IM-free DR. One way to do this is to view the first modulator as an optical pre-distorter for the second modulator. An advantage that such an optical pre-distorter has over its electronic counterpart is that changes in external factors such as temperature, aging, etc. do not change the *shape* of the pre-distorter transfer function; they only change the bias voltage at which a particular pre-distortion occurs.

The transfer function of the dual series modulator can be expressed as (Betts, 1998):

$$h(\phi_1, \phi_2, r) = \tfrac{1}{4}[1 + \cos(\phi_1 + m) + \cos(\phi_2 + rm)]$$
$$+ \tfrac{1}{8}[\cos(\phi_1 + \phi_2 + (1 + r)m) + \cos(\phi_1 - \phi_2 + (1 - r)m)],$$

(6.35)

where ϕ_1 and ϕ_2 are the bias points of the first and second modulator, respectively, and r is the relative phase of the RF voltage on the second modulator relative to the first modulator.

There are three degrees of freedom in (6.35) but we only have one constraint: minimization of the third-order distortion. Consequently we are free to impose two additional constraints. For the practical reasons cited above, we select as one of the constraints $r = 1$ so that a common set of electrodes may be used for both modulators. For the other additional constraint we choose $\phi_1 = \phi_2$; i.e. we choose to tie the two biases together, which means we need only one bias controller. With these constraints the third-order-only, dual series, linearized Mach–Zehnder modulator presents the same interface as a single Mach–Zehnder modulator.

Applying the above constraints to (6.35) we obtain the following expressions for its first three derivatives:

$$\frac{\partial h}{\partial V_M} = -\tfrac{1}{2}\left[\sin(\phi_1 + \pi m) + \tfrac{1}{2}\sin(2\phi_1 + 2\pi m)\right],$$

(6.36a)

$$\frac{\partial^2 h}{\partial V_M^2} = -\tfrac{1}{2}\left[\cos(\phi_1 + \pi m) + \cos(2\phi_1 + 2\pi m)\right],$$

(6.36b)

$$\frac{\partial^3 h}{\partial V_M^3} = \tfrac{1}{2}\left[\sin(\phi_1 + \pi m) + 2\sin(2\phi_1 + 2\pi m)\right].$$

(6.36c)

Figure 6.22 is a plot of (6.36a, b and c) each as a function of the bias ϕ_1. From Fig. 6.22 we see that the addition of a second Mach–Zehnder extends the linear portion of the transfer function near the quadrature bias point that is often used for analog modulation of a single Mach–Zehnder. For the constraints given, the bias for minimum third-order distortion is 104.48 degrees; i.e. there is now a useful minimum in the third-order distortion because it is not located at a zero of the fundamental.

It is important to note that at this bias point the third-order distortion is *minimized*, but *not* zero. To appreciate the reason for this distinction we need to examine in more detail the contributions to third-order distortion, that is, signals at $2f_1 - f_2$ or $2f_2 - f_1$. As one might surmise from the name, the dominant contributor to the third-order arises from the cubic term in the Taylor series for the device. However,

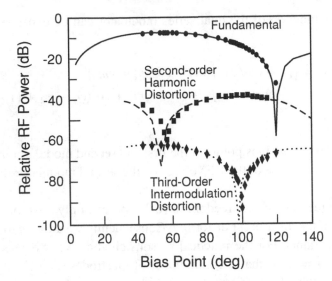

Figure 6.22 Calculated and measured plots of fundamental, second- and third-order distortion vs. bias voltage for the third-order-only, dual series, linearized Mach–Zehnder modulator (Betts, 1998). (Reprinted with permission from the author.)

it turns out that this is not the only contributor to third-order distortion – other higher-power odd-order terms, such as the fifth power term, also contribute to the third-order distortion, as well as to the fifth-order distortion. The bias point that was calculated above is the one that sets to zero the distortion arising from the cubic power term. Thus this bias sets most, but not all, of the third-order distortion to zero.

The bias for minimum third-order at $\phi_1 = 104.48$ degrees is near the second-order null at $\phi_1 = 90$ degrees. But since these two nulls do not occur at the same bias voltage, the second- and third-order distortions cannot be nulled simultaneously. Consequently this modulator cannot be used in links requiring linearization over more than an octave of bandwidth because of the second-order distortion that is present.

Betts (1994) has measured the IM-free DR for single and dual series modulators in a link under otherwise identical conditions. The IM-free DR for the single Mach–Zehnder link was 109 dB Hz$^{2/3}$ whereas the IM-free DR of the dual series link was 132 dB Hz$^{4/5}$. The improvement of 23 dB in IM-free DR is about the same as the improvements predicted by Bridges and Schaffner (1995) for the multi-electrode directional coupler modulator and demonstrated by Wilson *et al.* (1998) for their electrical pre-distortion of an EA modulator.

Figure 6.22 also plots the measured fundamental, second- and third-order distortion vs. bias voltage for the modulator as generally discussed above (Betts, 1998).

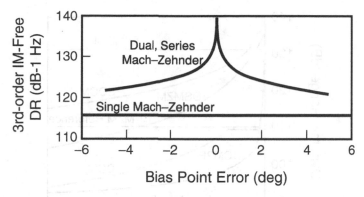

Figure 6.23 Plot of the IM-free DR vs. bias point error for the dual series Mach–Zehnder modulator (Betts, 1994, Fig. 7. © 1994 IEEE, reprinted with permission).

The experimental device involved some optimizations that moved the third-order minimum away from the value calculated above.

As we did with the pre-distortion of an EA modulator, we wish to explore the sensitivity of the IM-free DR improvement to changes in the bias. Figure 6.23 is a plot of the IM-free DR vs. bias for both the dual series and single Mach–Zehnder modulators. As we would expect from a process that is based on cancellation, the improvement degrades rapidly as one moves away from the cancellation condition. Thus with only a few tenths of a degree of bias error, the IM-free DR has dropped 10 dB. As the bias error increases, the reduction in IM-free DR becomes less severe. Consequently there is a range of several degrees over which the IM-free DR is better than for a single Mach–Zehnder. For very large bias errors, the IM-free DR of the linearized modulator can drop below that of a single Mach–Zehnder, at least for some design configurations (Betts, 1998).

It is important to keep in mind that Fig. 6.23 is shown for a bandwidth of 1 Hz. For practical systems the bandwidth will be greater than 1 Hz, and consequently the reduction with bias error will not be as strong a function of bias error. However, the maximum linearization improvement is also not as great.

Up to this point we have not included any specific reference to the frequency range over which the increased IM-free DR is obtained. As was discussed in Chapter 4, for high frequency operation, a traveling wave modulator with velocity match between the modulating and optical fields is the desired configuration. Cummings and Bridges (1998) have used computer simulations to model the IM-free DR vs. frequency for the standard (unlinearized) Mach–Zehnder and directional coupler modulators, as well as several of their linearized relatives. The results of their analysis are shown in Fig. 6.24. These curves apply to a specific length (10 mm) lithium niobate modulator with no velocity matching between the optical waveguides and the traveling wave electrodes, $n_{electrodes} - n_{optical} = \Delta n = 1.8$.

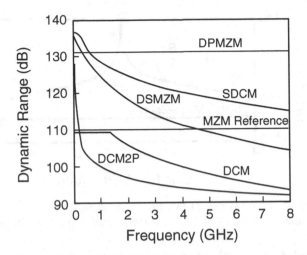

Figure 6.24 Plot of the IM-free DR vs. frequency for the single (unlinearized) Mach–Zehnder modulator (MZM reference) as well as two of its linearized relatives: the dual parallel (DPMZM) and dual series (DSMZM). Also shown is the IM-free DR for the standard (unlinearized) directional coupler modulator (DCM) as well as the two (SDCM) and three electrode (DCM2P) linearized directional coupler modulators (Cummings and Bridges, 1998, Fig. 10. © 1998 IEEE, reprinted with permission).

The frequency axis may be scaled for modulators with varying degrees of velocity match, Δn, or other lengths relative to 10 mm.

As was discussed in Section 6.3.2, the IM-free DR of the standard Mach–Zehnder modulator is ideally independent of bandwidth and the simulation results in Fig. 6.24 provide additional confirmation of this fact. The IM-free DR of the dual parallel Mach–Zehnder is also independent of frequency, since the linearization is obtained by incoherent addition of the parallel Mach–Zehnder outputs on the photodetector.

The results in Fig. 6.24 suggest that the IM-free DR of the dual series Mach–Zehnder modulator falls off significantly as the frequency increases. It turns out that this is true if there is a time delay between the modulation paths for each modulator, because the curves in Fig. 6.24 assumed that a single velocity-mismatched, traveling wave electrode drove both Mach–Zehnders. However, as pointed out by Betts (1998), one can eliminate the frequency dependence if the two modulators in the dual series Mach–Zehnder are driven such that the time delay of the path from the RF input through the first modulator to the optical output is the same as the time delay from the RF input through the second modulator to the optical output. This case was demonstrated experimentally by Betts and O'Donnell (1996).

Figure 6.24 also makes clear that the most rapid reduction in linearization with frequency occurs in the multi-electrode linearized version of the directional coupler modulator. The reason for this can be traced to the fundamentally different way the intensity modulation is produced in the directional coupler vs. the Mach–Zehnder. In the Mach–Zehnder, the phase modulation is distributed along the electrodes, but the intensity modulation occurs at a single point – where the two arms of the Mach–Zehnder recombine. On the other hand, in the directional coupler, the intensity modulation – and the resultant cancellation of intermodulation products – is distributed along the electrodes. Therefore linearization in a directional coupler imposes a tighter constraint on the degree of velocity match required to maintain a given level of linearization than does the bandwidth constraint.

Appendix 6.1 Non-linear distortion rate equation model for diode lasers

Recall from Appendix 2.1 that the macroscopic behavior of a semiconductor laser could be described by the following pair of equations that describe the interaction between the carrier density and photon density:

$$\frac{dn_U}{dt} = \frac{\eta_i i_L}{q V_E} - \frac{n_U}{\tau_U} - r_{SP} - v_g g(N_U, N_P)n_P, \tag{A6.1}$$

$$\frac{dn_P}{dt} = \Gamma v_g g(N_U, N_P)n_P + \Gamma \beta_{sp} r_{sp} - \frac{n_P}{\tau_P}. \tag{A6.2}$$

We see from the last term in (A6.1) and the first term in (A6.2) that fundamentally the rate equations are a pair of *non*-linear differential equations, primarily due to the stimulated emission terms, which are proportional to the *product* of the photon density and gain, which is in turn dependent on the carrier density. The spontaneous recombination terms are also not linear in the carrier density, but above threshold these contributions are negligibly small, especially at frequencies beyond a few megahertz.

In Appendix 4.1, we wanted to derive an expression for the small-signal frequency response of a diode laser above threshold. To do so we neglected any contributions of non-radiative recombination and spontaneous emission to the dynamics and linearized the rate equations by explicitly neglecting any terms in the rate equations that would produce non-linear distortion. For example we neglected terms containing the *product* of incremental variables.

In this appendix we keep the dominant product terms, which we will see are due to first-order variations in the gain multiplied by the variation in photon density, so that we can derive expressions for the non-linear distortion of diode lasers. We also continue to neglect dynamical contributions of non-radiative recombination and spontaneous emission with the consequence that the expressions we derive are only valid for frequencies near the relaxation resonance. This is because the cancellation of various distortion mechanisms at low frequencies is disrupted by

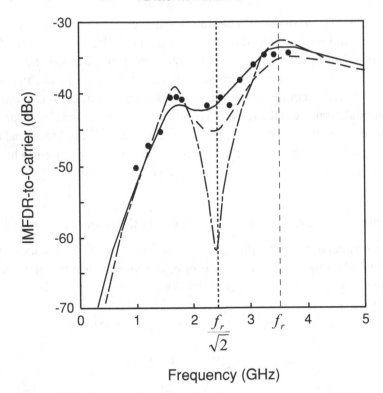

Figure A6.1 Plot of third-order intermodulation distortion vs. frequency; dash-dot curve Iannone and Darcie; dash curve Helms; solid curve Wang *et al.*; dots experimental data of Helms. (Helms, 1991; Wang *et al.*, 1993, Fig. 1; Iannone, P. and Darcie, T. 1987. Multichannel intermodulation distortion in high-speed GaInAsP lasers, *Electron. Lett.*, **23**, 1361–2. Reprinted with permission.

the dynamical contributions from non-radiative recombination and spontaneous emission. As we will see at the end of this appendix, without these terms the distortion predicted by this approach is much lower than allowed by a quasi-steady state analysis from the power series expansion of the dc *L–I* curve.

A review of the literature on the solutions to the non-linear rate equations reveals myriad different expressions for the third-order intermodulation distortion. In general the distinction among these expressions is the degree of simplifying approximations made in their derivation. As one might expect, the more terms that are kept, for example non-radiative recombination and spontaneous emission terms, the more complex the resulting formulae, but the more accurate the results. Figure A6.1 gives an example of this point. The "simple" formula we will present for third-order intermodulation distortion is plotted as the dash-dot curve in this figure. Note that when compared to the experimental data, this formula exaggerates the amount of distortion reduction between the peaks at the relaxation resonance frequency and at half the relaxation resonance frequency. The figure also contains the theoretical predictions of two more complex formulae, one by Helms (1991) and one by Wang *et al.* (1993), which increasingly improve the fit to the data.

Because the derivation of these expressions is not usually explicitly illustrated, it is difficult to see the consequences of these various approximations on the solutions. These issues will be addressed in what follows where we point out the deficiencies in using the approximations outlined in the previous paragraph. Once the reader appreciates the issues and the approach, it should prove easier to work with the more complex equations.

In the remainder of this appendix then, we want to outline the relatively simple procedure for calculating the distortion response of the laser that appeared in Chapter 6. The approach described here was first published by Lau and Yariv (1984) using an analysis that is similar to the one used in calculating the non-linear polarization response in non-linear optics. The updated and expanded version presented here is based on the work of Lee (2001). It is based on two major assumptions.

1. The first assumption is that the dominant response of the system is *linear* and the non-linearities serve only to generate harmonics and intermodulation products. This is reasonable because the linear response (low order) is much stronger than the non-linear response (higher order) in the small-signal limit. We used this assumption when deriving the small signal frequency response in Appendix 4.1.
2. The other assumption is that higher order harmonics and intermodulation products are orders of magnitude smaller than lower order harmonics and intermodulation products. This is reasonable based on the results from the power series analysis at low frequencies, the Taylor expansion of the stimulated emission terms, as well as our observations of the output spectrum of a modulated laser. This assumption allows us to calculate the magnitude and phase of the harmonics and intermodulation products based *solely* on the terms of lower order. For example, if we want to calculate the second harmonic distortion at 2ω, we only consider the term involving $e^{j\omega t}e^{j\omega t}$ and ignore the contribution from $e^{j3\omega t}e^{j\omega t}$ and $e^{j5\omega t}e^{j3\omega t}$, etc. because these higher order terms make negligible contributions to the second-order distortion.

These two assumptions allow the fundamental and distortion products to be solved in closed form in a hierarchical manner starting from the linear response, proceeding to the second-order harmonics and intermodulation distortions, which are in terms of the linear response, then to the third-order distortions, which are in terms of the linear response and the second-order responses, and so on.

We take as our starting point the expressions for the small-signal rate equations derived in Appendix 4.1, but before the linearization step, which we repeat below for convenience:

$$\frac{dn_u}{dt} = -v_g a_N N_P n_u - (v_g g(N_U, N_P) - v_g a_P N_P) n_p$$

$$+ \frac{\eta_i i_\ell}{q V_E} - v_g a_N n_u n_p + v_g a_P n_p^2, \tag{A4.5a}$$

$$\frac{dn_p}{dt} = \Gamma v_g a_N N_P n_u + \left(\Gamma v_g g(N_U, N_P) - \frac{1}{\tau_p} - \Gamma v_g a_P N_P \right) n_p$$

$$+ \Gamma v_g a_N n_u n_p - \Gamma v_g a_P n_p^2. \tag{A4.5b}$$

It is worth noting that we have neglected all distortion driving terms involving higher order derivatives of the gain because for typical gain parameter values the dominant non-linear driving terms are of the form $v_g a_N n_u n_p$ and $v_g a_P n_p^2$ shown in (A4.5). This can be seen by doing a rough order of magnitude estimate of the non-linear terms involving higher order derivatives. Equations (A4.5a,b) can be simplified into the following system of equations,

$$\frac{d}{dt} \begin{bmatrix} n_u \\ n_p \end{bmatrix} = \begin{bmatrix} -v_g a_N N_P & -v_g g_0 + v_g a_P N_P \\ \Gamma v_g a_N N_P & -\Gamma v_g a_P N_P \end{bmatrix} \begin{bmatrix} n_u \\ n_p \end{bmatrix} + \begin{bmatrix} \frac{\eta_i i_\ell}{q V_E} \\ 0 \end{bmatrix} + \cdots$$

$$+ \underbrace{\begin{bmatrix} -v_g a_N n_u n_p \\ \Gamma v_g a_N n_u n_p \end{bmatrix}}_{\text{intrinsic}} + \underbrace{\begin{bmatrix} v_g a_P n_p^2 \\ -\Gamma v_g a_P n_p^2 \end{bmatrix}}_{\text{gain compression}}, \tag{A6.3}$$

where the first set of non-linear driving terms has been named the intrinsic non-linearity and the second the gain compression non-linearity. Using more compact notation, we have

$$\frac{d}{dt} \begin{bmatrix} n_u \\ n_p \end{bmatrix} = \begin{bmatrix} -\gamma_{NN} & -\gamma_{NP} \\ \gamma_{PN} & -\gamma_{PP} \end{bmatrix} \begin{bmatrix} n_u \\ n_p \end{bmatrix} + \begin{bmatrix} \frac{\eta_i i_\ell}{q V_E} \\ 0 \end{bmatrix} + \cdots$$

$$+ \begin{bmatrix} -v_g a_N n_u n_p \\ \Gamma v_g a_N n_u n_p \end{bmatrix} + \begin{bmatrix} v_g a_P n_p^2 \\ -\Gamma v_g a_P n_p^2 \end{bmatrix}. \tag{A6.4}$$

A comment on our neglect of non-radiative recombination and spontaneous emission for this appendix is in order here. The two matrix elements most affected by this approximation are γ_{PP} and γ_{NN}. In going from (A4.5) to (A6.3), we have approximated $\Gamma v_g g - 1/\tau_p = 0$ but from (A2.1.16a), we see that $\Gamma v_g g - 1/\tau_p = -\Gamma \beta r_{sp}/N_P$, which would make $\gamma_{PP} = \Gamma v_g a_P N_P + \Gamma \beta r_{sp}/N_P$ instead of $\gamma_{PP} = \Gamma v_g a_P N_P$. Similarly, in the derivation of (A6.3), had we kept the differential non-radiative recombination and spontaneous emission terms, we would have arrived at $\gamma_{NN} = v_g a_N N_P + 1/\tau_n'$ instead of $\gamma_{NN} = v_g a_N N_P$. These omissions, as we will see later, have consequences on the calculated low frequency distortion.

We now have the form of the rate equations that we need to calculate distortion. An outline of the procedure for calculating the distortion is as follows. First, we choose an appropriate test solution for the small-signal carrier density and photon density response that contains the fundamental frequencies, harmonics, and intermodulation distortion products of interest. This test solution is substituted into (A6.4), and after multiplying out the products in the non-linear driving terms, we collect terms with the same frequency into separate systems of equations. Then, using the assumptions outlined above, we eliminate the higher order driving terms and solve the resulting systems of equations in a hierarchical manner, starting with the linear response and then proceeding to the second-order distortions and then the third.

To illustrate the process, we first calculate the second *harmonic* distortion (single fundamental driving frequency) and then, using these results, continue with the calculation of the third-order, *intermodulation* distortion (two fundamental driving frequencies), which is an important distortion parameter in many applications.

The appropriate test solution for calculating the second harmonic distortion is

$$n_u(t) = n_{u1}\, e^{j\omega t} + n_{u2}\, e^{j2\omega t}, \qquad n_p(t) = n_{p1}\, e^{j\omega t} + n_{p2}\, e^{j2\omega t}, \qquad i_1(t) = i_1\, e^{j\omega t}.$$

$$(A6.5)$$

After substituting (A6.5) into (A6.4) and collecting terms with the same frequency, we use the two assumptions listed above and separate the following *two* systems of equations, one for the fundamental response (A6.6), and the other for the second harmonic response (A6.10).

For the linear response, which we have already done in Appendix 4.1 but repeat here for convenience, the non-linear driving terms do not appear because $n_u n_p$ does not contribute at the fundamental frequency.

$$\begin{bmatrix} j\omega + \gamma_{NN} & +\gamma_{NP} \\ -\gamma_{PN} & j\omega + \gamma_{PP} \end{bmatrix} \begin{bmatrix} n_{u1} \\ n_{p1} \end{bmatrix} = \begin{bmatrix} \frac{\eta_i i_\ell}{q V_E} \\ 0 \end{bmatrix}.$$

$$(A6.6)$$

This is exactly the same system of equations we used to calculate the linear small-signal response in Appendix 4.1. To first order, it is not influenced at all by the higher order harmonics. The solution to the above system is obtained by using the formula for the inverse of a 2×2 matrix:

$$\begin{bmatrix} A & B \\ C & D \end{bmatrix}^{-1} = \frac{1}{AD - BC} \begin{bmatrix} D & -B \\ -C & A \end{bmatrix},$$

$$(A6.7)$$

which for the system matrix in (A6.6) is

$$\begin{bmatrix} j\omega + \gamma_{NN} & +\gamma_{NP} \\ -\gamma_{PN} & j\omega + \gamma_{PP} \end{bmatrix}^{-1} = \frac{H(\omega)}{\omega_r^2} \begin{bmatrix} j\omega + \gamma_{PP} & -\gamma_{NP} \\ +\gamma_{PN} & j\omega + \gamma_{NN} \end{bmatrix}.$$

$$(A6.8)$$

This gives the familiar results for the linear, small-signal carrier and photon density responses at the fundamental frequency ω:

$$n_{u1} = (j\omega + \gamma_{PP})\frac{\eta_i i_1}{q V_E}\frac{H(\omega)}{\omega_r^2},$$

$$n_{p1} = \gamma_{PN}\frac{\eta_i i_1}{q V_E}\frac{H(\omega)}{\omega_r^2}, \tag{A6.9}$$

where $H(\omega)/\omega_r^2$ is the reciprocal of the determinant of the system matrix.

For the second harmonic response, we again apply the two assumptions listed above. In this case the only driving terms are due to the product of the first-order response since $n_{u1}\,\mathrm{e}^{j\omega t}\,n_{p1}\,\mathrm{e}^{j\omega t} = \frac{1}{2}\,n_{u1}n_{p1}\,\mathrm{e}^{j2\omega t} + \frac{1}{2}\,n_{u1}n_{p1}$ and similarly for the n_{p1}^2 terms. Taking the terms with the 2ω dependence, the system of equations for the second harmonic response becomes

$$\begin{bmatrix} j2\omega + \gamma_{NN} & \gamma_{NP} \\ -\gamma_{PN} & j2\omega + \gamma_{PP} \end{bmatrix}\begin{bmatrix} n_{u2} \\ n_{p2} \end{bmatrix} = \frac{1}{2}\begin{bmatrix} -v_g a_N n_{u1}n_{p1} \\ \Gamma v_g a_N n_{u1}n_{p1} \end{bmatrix} + \frac{1}{2}\begin{bmatrix} v_g a_P n_{p1}^2 \\ -\Gamma v_g a_P n_{p1}^2 \end{bmatrix}.$$

$$\tag{A6.10}$$

Using A6.7, the solution to (A6.10) is

$$n_{u2} = -v_g a_N[(j2\omega + \gamma_{PP}) + \gamma_{NP}\Gamma]\frac{H(2\omega)}{\omega_r^2}\frac{n_{u1}n_{p1}}{2} + \cdots$$

$$+ v_g a_P[(j2\omega + \gamma_{PP}) + \gamma_{NP}\Gamma]\frac{H(2\omega)}{\omega_r^2}\frac{n_{p1}^2}{2},$$

$$n_{p2} = v_g a_N[\Gamma(j2\omega + \gamma_{NN}) - \gamma_{PN}]\frac{H(2\omega)}{\omega_r^2}\frac{n_{u1}n_{p1}}{2} + \cdots$$

$$- v_g a_P[\Gamma(j2\omega + \gamma_{NN}) - \gamma_{PN}]\frac{H(2\omega)}{\omega_r^2}\frac{n_{p1}^2}{2}. \tag{A6.11a,b}$$

We can use the following two relations to simplify (A6.11): $\Gamma\gamma_{NN} - \gamma_{PN} = 0$ when non-radiative recombination and spontaneous emission are neglected, and $\Gamma\gamma_{NP} + \gamma_{PP} = 1/\tau_p$, which can be seen from (A6.4) and the subsequent discussion. Using these relations to simplify (A6.11), we have

$$n_{u2} = -v_g(a_N n_{u1} - a_P n_{p1})\left(j2\omega + \frac{1}{\tau_p}\right)\frac{H(2\omega)}{\omega_r^2}\frac{n_{p1}}{2},$$

$$n_{p2} = v_g(a_N n_{u1} - a_P n_{p1})\Gamma(j2\omega) + \frac{H(2\omega)}{\omega_r^2}\frac{n_{p1}}{2}. \tag{A6.12a,b}$$

When ignoring spontaneous and non-radiative recombination, the low frequency intrinsic distortion, which is proportional to a_N, and the gain compression distortion, which is proportional to a_P, cancel as shown below:

$$(a_N n_{u1} - a_P n_{p1}) = [a_N(j\omega + \gamma_{PP}) - a_P \gamma_{PN}]\frac{\eta_i i_1}{q V_E}\frac{H(\omega)}{\omega_r^2}$$

$$= a_N j\omega \frac{\eta_i i_1}{q V_E}\frac{H(\omega)}{\omega_r^2}. \qquad (A6.13)$$

Substituting (A6.13) into (A6.12) results in expressions for the carrier and photon densities at the second harmonic:

$$n_{u2} = \left(\omega^2 - \frac{j\omega}{2\tau_P}\right)v_g a_N \frac{\eta_i i_1}{q V_E}\frac{H(\omega)}{\omega_r^2}\frac{H(2\omega)}{\omega_r^2}\left[\gamma_{PN}\frac{\eta_i i_1}{q V_E}\frac{H(\omega)}{\omega_r^2}\right],$$

$$n_{p2} = -\omega^2 \Gamma v_g a_N \frac{\eta_i i_1}{q V_E}\frac{H(\omega)}{\omega_r^2}\frac{H(2\omega)}{\omega_r^2}\left[\gamma_{PN}\frac{\eta_i i_1}{q V_E}\frac{H(\omega)}{\omega_r^2}\right], \qquad (A6.14a,b)$$

where the term in brackets is the response of the photon density at the fundamental.

Including the effects of non-radiative recombination and spontaneous emission would have modified the expression for n_{p2} in (A6.12) and (A6.13). In (A6.12) the $(j2\omega)$ term would have been replaced by $(j2\omega + 1/\tau_n')$ and in (A6.13) the $(j\omega)$ term would have been replaced by $(j\omega + \Gamma B r_{sp}/N_P)$. The impact of these two changes would have been to shift the zeros of the frequency response away from dc. Thus, we see that non-radiative recombination and spontaneous emission only affect the low frequency portion of the distortion response.

Dividing (A6.14) by the fundamental response, (A6.9), gives the final result for the second harmonic distortion relative to the fundamental (which is the expression arrived at by Darcie *et al.* (1985)):

$$\frac{n_{p2}}{n_{p1}} = -\Gamma v_g a_N \omega^2 \frac{\eta_i I_1}{q V_E}\frac{H(\omega)}{\omega_r^2}\frac{H(2\omega)}{\omega_r^2}$$

$$= -(OMD)\omega^2 \frac{H(2\omega)}{\omega_r^2}, \qquad (A6.15)$$

where the optical modulation depth (*OMD*) is defined as n_{p1}/N_P.

To convert the photon densities to measurable parameters, such as RF power, requires a number of steps (which we have done before in previous appendices): convert photon density to output optical power, then convert optical power to laser current and finally convert current to RF power. However, in this case these conversions affect both the numerator and denominator of (A6.15) and so in this case cancel out. Thus the square of (A6.15) is also the expression for the ratio of second-to-fundamental RF power.

We are now in a position to derive the expression for the third-order *intermodulation* distortion. To do this, we first need to assume a form of the carrier and photon densities that has two fundamentals, rather than the single fundamental we assumed when calculating harmonic distortion. For the photon density, we use

$$
\begin{aligned}
n_p = {} & n_{p1}\, e^{j\omega_1 t} + n_{p1}\, e^{j\omega_2 t} + n_{p21}\, e^{j(\omega_1+\omega_2)t} + n_{p22}\, e^{j(\omega_1-\omega_2)t} + n_{p23}\, e^{j2\omega_1 t} \\
& + n_{p23}\, e^{j2\omega_2 t} + n_{p31}\, e^{j(2\omega_1-\omega_2)t} + n_{p31}\, e^{j(2\omega_2-\omega_1)t} + n_{p32}\, e^{j(2\omega_1+\omega_2)t} \\
& + n_{p32}\, e^{j(2\omega_2+\omega_1)t} + n_{p33}\, e^{j3\omega_1 t} + n_{p34}\, e^{j3\omega_2 t},
\end{aligned}
\tag{A6.16}
$$

where we have assumed the frequencies ω_1 and ω_2 are very close together and assigned the same variable name to frequency components with the same magnitude. The form for the carrier density is identical except for the subscript u instead of p.

We are interested in calculating the third-order intermodulation distortion, n_{p31}/n_{p1}, due to the non-linear driving terms, $n_u n_p$ and n_p^2. Following the same procedure for deriving the second harmonic distortion and using the assumptions listed at the beginning of the appendix, we find that for the $n_u n_p$ term, the generators of the $(2\omega_1 - \omega_2)$ terms are $1/[2(n_{u1}^* n_{p23} + n_{p1}^* n_{u23})]$ and for the n_p^2 term, the generators are $(n_{p1}^* n_{p23})$. Note the complex conjugates that appear, which correctly account for the $-\omega_2$ component of the intermodulation distortion. Also, the $(n_{u22} n_{p1})$ terms are ignored because the response at $\omega_1 - \omega_2$ is very small when we ignore non-radiative recombination and spontaneous emission.

After substituting the two tone test solution, (A6.16), into (A6.4) and collecting terms with the $(2\omega_1 - \omega_2)$ dependence we arrive at the following system of equations for the intermodulation distortion, n_{p23}:

$$
\begin{bmatrix} j\omega + \gamma_{NN} & \gamma_{NP} \\ -\gamma_{PN} & j\omega + \gamma_{PP} \end{bmatrix}
\begin{bmatrix} n_{u31} \\ n_{p31} \end{bmatrix}
= \frac{1}{2}
\begin{bmatrix} -v_g a_N (n_{u1}^* n_{p23} + n_{p1}^* n_{u23}) \\ \Gamma v_g a_N (n_{u1}^* n_{p23} + n_{p1}^* n_{u23}) \end{bmatrix} + \cdots
$$
$$
+ \begin{bmatrix} v_g a_P n_{p1}^* n_{p23} \\ -\Gamma v_g a_P n_{p1}^* n_{p23} \end{bmatrix}.
\tag{A6.17}
$$

Using (A6.7) and following steps similar to the progression from (A6.10) to (A6.12), the solution for the expression for the third-order intermodulation distortion is

$$
n_{p31} = v_g [(a_N n_{u1}^* n_{p23} - a_P n_{p1}^* n_{p23}) + a_N n_{p1}^* n_{u23} - a_P n_{p1}^* n_{p23}] \frac{\Gamma}{2}(j\omega)\frac{H(\omega)}{\omega_r^2},
\tag{A6.18}
$$

which, after substituting the simplification of (A6.13) and collecting terms, gives

$$
\begin{aligned}
n_{p31} &= v_g a_N \left[-j\omega \frac{\eta_1 i_1^*}{q V_E} \frac{H(\omega)^*}{\omega_r^2} n_{p23} + n_{p1}^* n_{u23} - \frac{a_P}{a_N} n_{p1}^* n_{p23} \right] \frac{\Gamma}{2}(j\omega)\frac{H(\omega)}{\omega_r^2} \\
&= v_g a_N \frac{\eta_1 i_1^*}{q V_E} \frac{H(\omega)^*}{\omega_r^2} \left[\gamma_{PN} n_{u23} - \left(j\omega + \frac{a_P}{a_N}\gamma_{PN} \right) n_{p23} \right] \frac{\Gamma}{2}(j\omega)\frac{H(\omega)}{\omega_r^2}.
\end{aligned}
\tag{A6.19}
$$

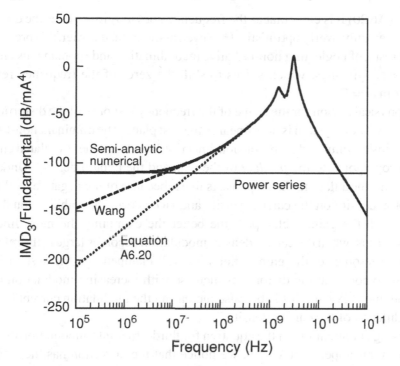

Figure A6.2 Plot of third-order intermodulation distortion relative to the fundamental vs. frequency for four different models. (Lee, 2001, reprinted with permission of the author.)

The forms of n_{p23} and n_{u23} are identical to the second harmonic distortion of (A6.14). Therefore after substituting (A6.13) into (A6.19), simplifying and dividing the result by (A6.9), we arrive at the following expression for the intermodulation distortion n_{p31}/n_{p1}:

$$\frac{n_{p31}}{n_{p1}} = \frac{1}{2}|OMD|^2\, H(\omega)H(2\omega)\left[\frac{1}{2}\left(\frac{\omega}{\omega_r}\right)^2 - \left(\frac{\omega}{\omega_r}\right)^4\right.$$
$$\left. + j\left(\frac{\omega}{\omega_r}\right)^3 \omega_r\tau_p\left(1 + \frac{\Gamma a_P}{a_N}\right)\right], \tag{A6.20}$$

where we have substituted $\omega_r^2 = v_g a_N N_P/\tau_p$ and $OMD = n_{p1}/N_P$.

Equation (A6.20) is plotted in Fig. A6.2, along with the more accurate equation from Wang *et al.* (1993) and the semi-analytic numerical solution of Lee (2001). The *y*-axis units are in electrical dB (optical squared) per mA^4 (input RF power squared). Note: the plot is for a constant modulation *current*, not a constant optical modulation *depth* as shown in some of the literature.

From (A6.20) it is clear that as the frequency goes to zero, so does the distortion, which is certainly overly optimistic. However, as mentioned earlier, this prediction is a direct result of neglecting non-radiative recombination and spontaneous emission in the laser dynamics, which serves to shift the zeros of the frequency response away from $\omega = 0$.

The physical reason for the shape of the frequency response of the distortions has to do with why the system is non-linear in the first place. The dominant non-linearity is the intrinsic stimulated emission non-linearity. Recall that the stimulated emission rate is proportional to the *product* of the gain and the photon density. Under very small-signal modulation, the response is linear because of the negative feedback of the photon density on the carrier density and vice versa. The stabilization depends on how well the gain is clamped: the better the clamping, the more linear the response. Thus, when the carrier density modulation becomes large at the relaxation oscillation resonance, the gain is not very well clamped and the system is non-linear. Also note that the distortions increase with increasing modulation current (note the units). Again, this is because increasing the modulation current increases the modulation of the carrier density.

Focusing our attention on the equation for third-order intermodulation distortion derived in this appendix (A6.20), we notice that the distortion has the resonance peaks discussed above and that the distortion falls off above and below resonance. Above resonance, the distortion falls off with increasing frequency because neither carrier density nor photon density can respond fast enough and the stimulated emission non-linearity is negligible.

Below resonance, the distortion also decreases with decreasing frequency. Initially, this roll-off is due to improved gain clamping (lower carrier density response) for lower frequencies. However, because of gain compression, the gain clamping is not perfect at dc and, using this line of reasoning, we expect some residual distortion instead of the continued roll-off. The continued roll-off is due to cancellations in the distortion response of carrier density and photon density (a cancellation of the intrinsic and gain compression non-linearities) at low frequencies, which were evident in going from (A6.11) to (A6.12) and in (A6.13). As mentioned above, this cancellation is not perfect if we include non-radiative recombination and spontaneous emission in the laser dynamics. In Wang's equation, he includes some of the effects of non-radiative and spontaneous emission, leading to an improved fit at low frequencies. However, it still suffers from a zero dc distortion due to some additional approximations where spontaneous emission terms were neglected.

At very low frequencies, the distortion does not go to zero. Physically, the distortion should approach the value given by a power series expansion of the *L–I* curve since it represents the dc transfer characteristic of the laser. This dc distortion model contains no dynamics and therefore is independent of frequency. Thus the *L–I* curve power series distortion model is accurate at low frequencies, not at high frequencies where it underestimates the distortion because it ignores the increase in distortion due to the reduction in the gain clamping. The perturbation analysis described in this appendix could have been extended to include all of the effects of non-radiative recombination and spontaneous emission, and the additional driving term $(n_{u22}n_{p1})$, to derive the low frequency distortion. However, such an effort is not justified because our laser model does not include some physical effects that are important at low frequencies, such as temperature variations and its effect on the gain.

References

Ackerman, E. 1999. Broadband linearization of a Mach–Zehnder electro-optic modulator, *IEEE Trans. Microwave Theory Tech.*, **47**, 2271–9.

Ackerman, E. I., Cox, C. H. III, Betts, G. E., Roussell, H. V., Ray, K. and O'Donnell, F. J. 1998. Input impedance conditions for minimizing the noise figure of an optical link, *IEEE Trans. Microwave Theory Tech.*, **46**, 2025–31.

Bertelsmeier, M. and Zschunke, W. 1984. Linearization of broadband optical transmission systems by adaptive predistortion, *Frequenz*, **38**, 206–12.

Betts, G. E. 1994. Linearized modulator for suboctave-bandpass optical analog links, *IEEE Trans. Microwave Theory Tech.*, **42**, 2642–9.

Betts, G. E. 1998. Personal communication.

Betts, G. E. and O'Donnell, F. J. 1996. Microwave analog optical links using suboctave linearized modulators, *IEEE Photon. Technol. Lett.*, **8**, 1273–5.

Betts, G. E., Walpita, L. M., Chang, W. S. C. and Mathis, R. F. 1986. On the linear dynamic range of integrated electrooptical modulators, *IEEE J. Quantum Electron.*, **22**, 1009–11.

Betts, G. E., Cox, C. H. III and Ray, K. G. 1990. 20 GHz optical analog link using an external modulator, *IEEE Photon. Technol. Lett.*, **2**, 923–5.

Betts, G. E., Johnson, L. M. and Cox, C. H. III 1991. Optimization of externally modulated analog optical links, *Devices for Optical Processing, Proc. SPIE*, **1562**, 281–302.

Betts, G. E., O'Donnell, F. J. and Ray, K. G. 1995. Sub-octave-bandwidth analog link using linearized reflective modulator, *PSAA-5 Proceedings*, pp. 269–99.

Bridges, W. B. 2001. Personal communication.

Bridges, W. B. and Schaffner, J. H. 1995. Distortion in linearized electrooptic modulators, *IEEE Trans. Microwave Theory Tech.*, **43**, 2184–97.

Brooks, J., Maurer, G. and Becker, R. 1993. Implementation and evaluation of a dual parallel linearization system for AM-SCM video transmission, *J. Lightwave Technol.*, **11**, 34.

Bulmer, C. H. and Burns, W. K. 1983. Linear interferometric modulators in Ti:LiNbO$_3$, *J. Lightwave Technol.*, **2**, 512–21.

Childs, R. B. and O'Byrne, V. A. 1990. Multichannel AM video transmission using a high-power Nd:YAG laser and linearized external modulator, *IEEE J. Selected Areas Commun.*, **8**, 1376–96.

Cummings, U. V. and Bridges, W. B. 1998. Bandwidth of linearized electro-optic modulators, *J. Lightwave Technol.*, **16**, 1482–90.

Darcie, T. E., Tucker, R.S. and Sullivan, G. J. 1985. Intermodulation and harmonic distortion in InGaAsP lasers, *Electron. Lett.*, **21**, 665–6. See also correction in vol. 22, p. 619.

Dentan, M. and de Cremoux, B. 1990. Numerical simulation of the nonlinear response of a p-i-n photodiode under high illumination, *J. Lightwave Technol.*, **8**, 1137–44.

deRidder, R. M. and Korotky, S. K. 1990. Feedforward compensation of integrated optic modulator distortion, *Proc. Optical Fiber Communications Conference*, paper WH5.

Esman, R. D. and Williams, K. J. 1990. Measurement of harmonic distortion in microwave photodetectors, *IEEE Photon. Technol. Lett.*, **2**, 502–4.

Farwell, M. L., Lin, Z. Q., Wooten, E. and Chang, W. S. C. 1991. An electrooptic intensity modulator with improved linearity, *IEEE Photon. Technol. Lett.*, **3**, 792–5.

Frankart *et al.* 1983. Analog transmission of TV-channels on optical fibers, with nonlinearities corrected by regulated feedforward, *Proc. European Conference on Optical Communications (ECOC)*, pp. 347–50.

Giboney, K. S., Rodwell, M. J. W. and Bowers, J. E. 1997. Traveling-wave photodetector theory, *IEEE Trans. Microwave Theory Tech.*, **45**, 1310–19.

Gray, P. E. and Searle, C. L. 1969. *Electronic Principles Physics, Models, and Circuits*, New York: John Wiley & Sons, Inc., Section 4.3.2.

Helms, J. 1991. Intermodulation and harmonic distortions of laser diodes with optical feedback, *J. Lightwave Technol.*, **9**, 1567–75.

Iannone, P. and Darcie, T. 1987. Multichannel intermodulation distortion in high-speed GaInAsP lasers, *Electron. Lett.*, **23**, 1361–2.

Jasmin, S., Vodjdani, N., Renaud, J.-C. and Enard, A. 1997. Diluted- and distributed-absorption microwave waveguide photodiodes for high efficiency and high power, *IEEE Trans. Microwave Theory Tech.*, **45**, 1337–41.

Johnson, L. M. and Roussell, H. V. 1988. Reduction of intermodulation distortion in interferometric optical modulators, *Opt. Lett.*, **13**, 928–30.

Kim, E. M., Tucker, M. E. and Cummings, S. L. 1989. Method for including CTBR, CSO and channel addition coefficient in multichannel AM fiber optic system models, *NCTA Technical Papers*, p. 238.

Kurazono, S., Iwasaki, K. and Kumagai, N. 1972. A new optical modulator consisting of coupled optical waveguides, *Electron. Comm. Jap.*, **55**, 103–9.

Lau, K. Y. and Yariv, A., 1984. Intermodulation distortion in a directly modulated semiconductor injection laser, *Appl. Phys. Lett.*, **45**, 1034–6.

Lee, H. 2001. *Direct Modulation of Multimode Vertical Cavity Surface Emitting Lasers*, Master of Science Thesis, MIT, Cambridge, MA.

Lin, L. Y., Wu, M. C., Itoh, T., Vang, T. A., Muller, R. E., Sivco, D. L. and Cho, A. Y. 1997. High-power high-speed photodetectors – Design, analysis and experimental demonstrations, *IEEE Trans. Microwave Theory Tech.*, **45**, 1320–31.

Liu, P., Li, B. and Trisno, Y. 1991. In search of a linear electrooptic amplitude modulator, *IEEE Photon. Technol. Lett.*, **3**, 144–6.

Martin, W. 1975. A new waveguide switch/modulator for integrated optics, *Appl. Phys. Lett.*, **26**, 562–4.

Nazarathy, M., Berger, J., Ley, A. J., Levi, I. M. and Kagan, Y. 1993. Progress in externally modulated AM CATV transmission systems, *J. Lightwave Technol.*, **11**, 82–105.

Ozeki, T. and Hara, E. H. 1976. Measurements of nonlinear distortion in photodiodes, *Electron. Lett.*, **12**, 80.

Petermann, K. 1988. *Laser Diode Modulation and Noise*, Dordrecht, The Netherlands: Kluwer Academic Publishers, Section 4.7.

Phillips, M. R. and Darcie, T. E. 1997. Lightwave analog video transmission. In *Optical Fiber Communications IIIA*, I. P. Kaminow and T. L. Koch, eds., San Diego, CA: Academic Press, Chapter 14.

Prince, J. L. 1998. Personal communication.

Roussell, H. V. 1995. Personal communication.

    2001. Personal communication.

Skeie, H. and Johnson, R. V. 1991. Linearization of electro-optic modulators by a cascade coupling of phase modulating electrodes, *Proc. SPIE*, **1583**, 153–64.

Straus, J. 1978. Linearized transmitters for analog fiber links, *Laser Focus*, October, 54–61.

Thomas, G. B. 1968. *Calculus and Analytical Geometry*, Reading, MA: Addison-Wesley Publishing Co.

Wang, J., Haldar, M. K. and Mendis, F. V. C., 1993. Formula for two-carrier third-order intermodulation distortion in semiconductor laser diodes, *Electron. Lett.*, **29**, 1341–3.

Welstand, R. B., Zhu, J. T., Chen, W. X., Yu, P. K. L. and Pappert, S. A. 1999. Combined Franz-Keldysh and quantum-confined Stark effect waveguide modulator for analog signal transmission, *J. Lightwave Technol.*, **17**, 497–502.

Williams, A. R., Kellner, A. L. and Yu, P. K. L. 1993. High frequency saturation measurements of an InGaAs/InP waveguide photodetector, *Electron. Lett.*, **29**, 1298–9.

Williams, K. J., Esman, R. D. and Dagenais, M. 1996. Nonlinearities in p-i-n microwave photodiodes, *J. Lightwave Technol.*, **14**, 84–96.

Wilson, G. C., Wood, T. H., Gans, M., Zyskind, J. L., Sulhoff, J. W., Johnson, J. E., Tanbun-Ek, T. and Morton, P. A. 1998. Predistortion of electroabsorption modulators for analog CATV systems at 1.55 μm, *J. Lightwave Technol.*, **15**, 1654–61.

WJ 1998. Preliminary data sheet for the AH22 High Dynamic Range Amplifier, WJ Wireless Products Group.

Yu, P. K. L. 1998. Personal communication.

7

Link design tradeoffs

7.1 Introduction

Up to this point we have discussed each of the primary measures of link performance – gain, bandwidth, noise figure and dynamic range – in as complete isolation from the other parameters as possible. While such an approach permitted us to focus on the various aspects of each parameter, it did miss the effects of interactions among the parameters to a large extent. Clearly when designing a link, one needs to take into account such interactions; in fact one might argue that maturity in link design is gauged by the link designer's ability to balance often conflicting requirements to meet a given combination of link parameters.

As one might expect, there are myriad potential interactions among link parameters. Therefore in this chapter we can only offer a sampling of these interactions. We begin by exploring interactions among the primary parameters of the intrinsic link. In general the best link designs usually result from attaining the required performance via optimization of the intrinsic link.

However, there are situations where despite a link designer's best efforts, the intrinsic link performance falls below the requirements. In some of these cases electronic pre- and/or post-amplification may be used to improve performance. Consequently we expand our interaction space to include a sampling of tradeoffs between amplifier and link parameters.

7.2 Tradeoffs among intrinsic link parameters

7.2.1 Direct modulation

7.2.1.1 Diode laser bias current

In Fig. 2.2 we saw that the slope efficiency of a directly modulated link is highest just above threshold and decreases as the bias current is increased above threshold – slowly at first and then more rapidly as the bias current is increased further. Thus,

Table 7.1 *Summary of key direct modulation*
link parameters and their optimum bias

Parameter	Optimum bias
Gain	Low
Bandwidth	High
Noise figure	Low to moderate
IM-free DR	Moderate

from a link gain viewpoint, we wish to operate the laser above but as close as possible to threshold.

Recall from (4.4) that the relaxation frequency, and hence the maximum modulation frequency, of a laser increases as the square root of the bias current above threshold. Thus broadband or high frequency narrow-band links require biasing the laser well above threshold. For example broad bandwidth modulation is typically achieved with the laser biased at ten times threshold. The increase in bandwidth with increasing bias current is theoretically unbounded, but practically limited due to factors such as the thermal dissipation of the laser mount. So from a bandwidth perspective, the further the bias current is above threshold the better.

From a noise figure perspective, Fig. 5.11 shows that at best the noise figure is approximately independent of laser bias above threshold. In cases of high RIN, however, the noise figure increases as the laser bias increases. Therefore from a noise figure viewpoint low to moderate bias currents are desirable.

In Fig. 6.8, we saw that both the second- and third-order intermodulation-free dynamic ranges (IM-free DRs) are relatively complicated functions of the laser bias above threshold. However, there is a broad maximum for both these measures of dynamic range that occurs well above threshold but below the onset of saturation.

The result of all these bias dependent trends is a classic design tradeoff as summarized in Table 7.1.

From the table we can see that some of the parameter vs. bias trends pull the bias current selection in diametrically opposite directions: low bias for maximum gain vs. high bias for maximum bandwidth. Other parameter vs. bias trends are more accommodating to a common choice of bias current: moderate bias current for minimum noise figure and maximum IM-free DR.

What approach should the link designer take in such a situation? Well, one way to deal with conflicting design constraints is to prioritize them. In the case under discussion here, little if anything can be done external to the laser to increase its bandwidth. Consequently the bias current needs to be chosen such that the laser can be modulated at the desired frequency.

With the bias current set by the bandwidth constraint, there is often the need to invoke some techniques to improve the gain. In narrow-band applications, bandpass impedance matching, as discussed in Section 4.3.2.2, can be used. Alternatively an amplifier can be used alone or in conjunction with the matching. In broadband applications an amplifier is probably the only choice. However, as we will see in Section 7.3, the introduction of amplifiers introduces its own set of tradeoffs.

We can quantify some of these tradeoffs by applying equations derived in earlier chapters. For example consider the tradeoff between noise figure and IM-free DR as a function of average laser bias current, which is often encountered in practice (Ackerman, 1998). Since the range of bias currents in which we are interested lies above threshold, we can explore these dependences by examining the average photodiode current – which can be approximated as being proportional to the bias current in this range. If we continue to assume that $T_{FF} = 1$, then $P_{O,D} = P_{L,O}$, and thus we can establish the desired relationship between laser and photodiode currents: $I_D = r_d P_{O,D} = r_d P_{L,O} = r_d s_l (I_L - I_T)$.

Recall from Chapter 5 that with direct modulation, the laser RIN is usually the dominant noise source. In this case the link noise figure in terms of optical power can be obtained by substituting (5.12) into (5.17):

$$NF_{RIN} = 10 \log \left(2 + \frac{\langle I_D \rangle^2 10^{\frac{RIN}{10}} R_{LOAD}}{2 s_l^2 r_d^2 k T} \right). \tag{7.1}$$

Equation (7.1) suggests that for low laser bias currents, the noise figure is constant, then increases quadratically as the laser bias current increases. This dependence is plotted in Fig. 7.1 assuming the parameters listed in Table 5.1, with the additional assumption that $RIN = -155$ dB/Hz.

To complete the quantitative view of the noise figure vs. dynamic range tradeoff, we need an expression for IM-free DR that contains an explicit dependence on the average laser bias current. For the purposes of this discussion we assume that the photodiode intermodulation distortion is negligible compared to the laser distortion. It is easy to show that the third-order intercept point of a laser with a strictly linear P vs. I curve above threshold increases as the square of the average laser current:

$$IP_3 = 10 \log \left(\frac{(I_L - I_T)^2 R_L}{2} \right). \tag{7.2}$$

This equation is useful for examining trends and tradeoffs, such as we are doing here. But the reader is cautioned against using (7.2) for making specific predictions, as experience has shown that (7.2) overestimates the IM-free DR by up to as much as 20 dB.

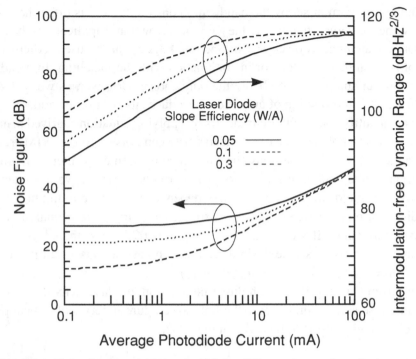

Figure 7.1 Noise figure and third-order IM-free DR vs. average photodiode current for a directly modulated link with laser slope efficiency as a parameter.

To express (7.2) in terms of average photodiode current, we substitute the expression just derived relating laser to photodiode current into (7.2) to obtain

$$IP_3 = 10 \log \left(\frac{I_D^2 R_{LOAD}}{2 s_\ell^2 r_d^2} \right). \tag{7.3}$$

By examining the geometry of a plot of output signal, noise and intermodulation distortion powers vs. input signal power, as we did in Section 6.2.2, it is also relatively easy to show that the third-order intercept point, IP_3, noise figure and third-order intermodulation-free dynamic range, IMF_3, are related by the following formula (which was first derived as (6.7) but is repeated below for convenience):

$$IMF_3 = \tfrac{2}{3}[IP_3 - (G + NF + kT\,\Delta f)]. \tag{7.4}$$

Recall from Chapter 5 that in a 1-Hz bandwidth and at the standard temperature of 290 K, the last term in (7.4) evaluates to approximately −204 dBW.

We now substitute expressions for intercept point, gain and noise figure, (7.3) (3.7) and (7.1), respectively, into (7.4) and plot the resulting expression in Fig. 7.1. Over the range of laser bias currents where the noise figure is essentially independent

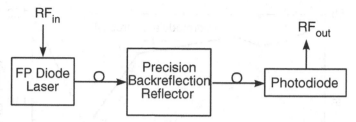

Figure 7.2 Block diagram of a Fabry–Perot (FP) laser-based directly modulated link for measuring the effects of optical back reflections on link noise figure and IM-free DR.

of this current, the IM-free DR increases as the square of laser current. Once the laser current reaches a level where the noise figure begins to increase quadratically with current, the IM-free DR becomes independent of laser bias.

The tradeoffs vs. laser bias current are actually more complex than this figure indicates. One of the complicating factors is that the RIN is not constant with laser bias, but typically decreases quadratically with bias above threshold.

7.2.1.2 Optical reflections

The high optical gain of a semiconductor laser is the key to many of its unique properties among lasers, such as its high modulation bandwidth. As was discussed in Chapter 2, a laser is formed by imbedding the gain medium in a cavity whose optical reflections make it optically resonant at the same wavelength as the gain. The introduction of additional optical reflections – from the laser-to-fiber coupling, from fiber connectors or from the photodiode – essentially forms additional optical cavities. This is why extraneous optical reflections in a link need to be minimized.

In a succinct set of experiments Roussell *et al.* (1997) clarified the deleterious effects of optical reflections on link noise figure and IM-free DR. In these experiments they used a simple direct modulation link that had one controlled optical reflection at the fiber-to-photodiode interface as shown in Fig. 7.2. Care was taken to suppress all extraneous optical reflections in the link so that the controlled optical reflection would dominate. The controlled optical reflection was introduced by fusing in one of a series of optic fibers whose ends had been angled to produce the desired optical reflection.

Figure 7.3 is a plot of noise figure and third-order, IM-free DR vs. optical reflection expressed in dB relative to a perfect reflector. As the data make clear, the optical reflections must be kept below −40 dB (i.e. less than 0.01%) to avoid having the optical reflections degrade the link performance. To put this amount of reflection in perspective, the reflection from a glass-to-air interface is about 4%. Therefore to reduce reflections from the fiber ends into the range where they have negligible

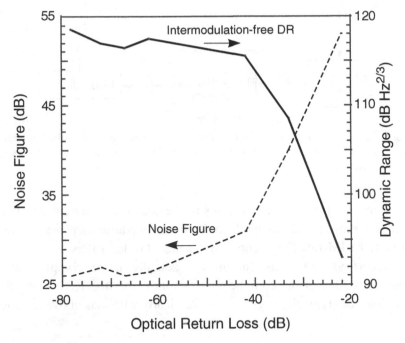

Figure 7.3 Noise figure and third-order IM-free DR vs. optical return loss for a directly modulated link (Roussell *et al.*, 1997, Fig. 3. © 1997 IEEE, reprinted with permission).

effects on the link performance requires some form of reflection suppression, such as angling or anti-reflection coating the fiber end faces.

Since reducing the reflections improves the link performance without introducing any deleterious side effects, one may ask: why not always include some form of reflection suppression? The answer is often simply the added cost of reflection suppression. Consequently reflection control is a cost–performance tradeoff, as opposed to the laser bias point selection, which is a performance–performance tradeoff.

7.2.2 External modulation

7.2.2.1 Modulator bias point

In the most common modulator under discussion here, a Mach–Zehnder fabricated in lithium niobate, the bandwidth is not a function of bias, so that variables can be eliminated from the tradeoff discussions for external modulators.

Throughout the discussions in the preceding chapters, we have assumed that the modulator was biased at or near the center of the range of its transfer function. As

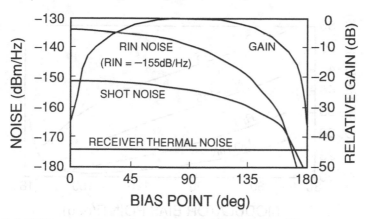

Figure 7.4 RIN power, shot noise power and relative link gain vs. Mach–Zehnder
bias point (Betts *et al.*, 1997, Fig. 14. © 1997 IEEE, reprinted with permission).

we saw in Section 6.3.2, this bias corresponds to the highest gain and minimum
second-order distortion.

However, it was noticed almost simultaneously by three groups (Ackerman
et al., 1993; Betts and O'Donnell, 1993; a more complete and generally accessible version was published by Betts *et al.*, 1997; and Farwell *et al.*, 1993) that a
lower noise figure can be obtained by biasing the modulator away from this bias
point.

Recall that all earlier link expressions for gain, noise figure, etc. that depend on
modulator incremental modulation efficiency were derived assuming a particular
bias point. Consequently, to examine the effects of bias point on these various link
parameters, we need to start with a new expression for the incremental modulation
efficiency. Betts and O'Donnell (1993) has shown that for a Mach–Zehnder modulator, the incremental modulation efficiency, (2.17), at quadrature bias, i.e. $\phi_B =
90°$, can be modified simply to accommodate any bias by multiplying (2.17) by
$\sin^2(\phi_B)$:

$$\frac{p_{m,o}^2}{p_{s,a}}(\phi_B) = \frac{s_{mz}^2}{R_S} \sin^2(\phi_B). \tag{7.5}$$

If follows immediately that the link gain, (3.10), as a function of modulator bias
becomes

$$g_{mzpd}(\phi_B) = s_{mz}^2 r_d^2 \sin^2(\phi_B). \tag{7.6}$$

This confirms the fact that was pointed out in Section 2.2.2.1 that the incremental
modulation efficiency, and hence the link gain, is maximum at quadrature bias.
Equation (7.6) is plotted vs. modulator bias angle in Fig. 7.4.

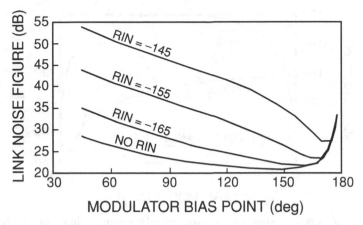

Figure 7.5 Link noise figure vs. Mach–Zehnder bias point with the CW laser RIN as a parameter (Betts *et al.*, 1997, Fig. 15. © 1997 IEEE, reprinted with permission).

Betts has also shown that a version of the link noise figure expression for quadrature bias, (5.32), that is applicable for any modulator bias angle is

$$NF(\phi_B) = 10\log\left\{2 + \frac{N_D^2 R_{\text{LOAD}}}{g_i kT}\left[\langle i_{\text{rin}}^2\rangle\left(\frac{(1+\cos(\phi_B))^2}{\sin^2(\phi_B)}\right)\right.\right.$$
$$\left.\left. + \langle i_{\text{sn}}^2\rangle\left(\frac{1+\cos(\phi_B)}{\sin^2(\phi_B)}\right) + \langle i_t^2\rangle\left(\frac{1}{\sin^2(\phi_B)}\right)\right]\right\}. \tag{7.7}$$

The $\sin^2$ term is symmetric about quadrature bias; however, the $1 + \cos(\phi_B)$ terms are smaller for biases above quadrature than for biases below. Thus the shot noise, and to a greater extent the RIN, contributions are attenuated by biasing the modulator above quadrature. This effect is also shown in Fig. 7.4 by the plots of the shot and RIN terms of (7.7).

Figure 7.5 is a plot of the noise figure as a function of modulator bias, (7.7), with the CW laser RIN as a parameter. This plot makes clear the noise figure advantage of biasing the modulator above quadrature. This plot also makes clear that there is no noise figure advantage to biasing the modulator too close to cutoff, i.e. $\phi_B = 180°$. The reason for this is that once the shot noise and RIN have been attenuated below the receiver noise, this latter noise now dominates the noise figure. Thus further increases in the bias angle do not decrease the noise, but they do decrease the gain.

What about the IM-free DR as a function of modulator bias? As was shown in Fig. 6.10, the second-order distortion for a Mach–Zehnder is zero at quadrature bias but increases symmetrically on either side of this bias point. Consequently, the second-order intermodulation suppression, which is the ratio of the second-order distortion to the fundamental, degrades on either side of quadrature bias. The

analytical form of the second-order suppression vs. bias can be obtained by taking
the ratio of (6.19b) to (6.19a):

$$\frac{p_{2nd}}{p_{fund}} \propto \left(\frac{\pi}{V_\pi}\right) \cot\left(\frac{\pi V_M}{V_\pi}\right). \tag{7.8}$$

Third-order distortion was zero when the modulator was biased for either maximum or minimum transmission, and increased in between. To determine the third-
order suppression, we take the ratio of (6.19c) to (6.19a):

$$\frac{p_{3nd}}{p_{fund}} \propto \left(\frac{\pi}{V_\pi}\right)^2, \tag{7.9}$$

which is *independent* of the modulator bias point. Therefore for biases above quadrature where the noise figure is decreasing, the third-order IM-free DR of a Mach–
Zehnder modulator *increases*. For shot noise limited links the improvement is only
3 dB; however, with RIN-dominated links, the improvement can be considerably
higher – until the receiver noise dominates.

The advantages of biasing a Mach–Zehnder above quadrature are useful only
in systems with bandwidth less than an octave where the increased second-order
distortion can be filtered out. Since many external modulation links use a solid
state laser for the CW source, the RIN is already negligible and consequently the
advantages of bias above quadrature are small. On the other hand, this bias technique
should permit the link designer to avoid the RIN penalties that almost invariably
accompany link designs that use a diode laser as the CW source.

7.2.2.2 Modulator incremental modulation efficiency

It is easy to get the impression from the material presented to this point that the
higher the incremental modulation efficiency the better: for example with increasing
incremental modulation efficiency, the gain increases and the noise figure decreases.
But do improvements such as these continue without bound as the modulation
efficiency increases?

Theoretically there is no maximum gain. Consequently for the gain the answer
to this question is yes – unlimited increases in modulation efficiency result, at least
theoretically, in unlimited increases in link gain. However, there is a theoretical
minimum noise figure, i.e. 0 dB, so it is clear from this consideration alone that
unlimited increases in modulation efficiency cannot result in unlimited reductions
in noise figure. In fact, as we saw in Section 5.5.1, the minimum noise figure for
some link configurations is not 0 dB but 3 dB as the modulation efficiency, and
hence the gain, go to infinity.

As we saw in Section 2.2.2.1 the principal parameters affecting the incremental
modulation efficiency of a Mach–Zehnder modulator are the average optical power

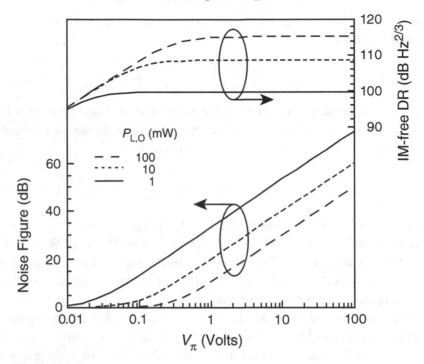

Figure 7.6 Noise figure and third-order IM-free DR vs. modulator V_π for an externally modulated link.

and the modulator half wave voltage, V_π. In Section 5.4.3 we derived an expression, (5.30), for the link noise figure as a function of these two parameters under shot-noise-dominated operation:

$$NF = 10\log\left(2 + \frac{2q\,R_{\text{LOAD}}}{\left(\frac{\pi R_S}{2V_\pi}\right)^2 T_{\text{FF}}P_{\text{I}}r_{\text{d}}kT}\right) \propto 10\log\left(2 + aV_\pi^2\right), \quad (7.10)$$

where a is a constant. For reasons that will become clear in the following, we have chosen to focus here on the V_π dependence and assume that the average optical power is fixed.

A plot of noise figure vs. V_π is shown in Fig. 7.6 for a representative set of link parameters. Intuitively what is happening here is that the increased modulation efficiency improves the link's ability to detect smaller and smaller signals. We are not always able to utilize this increased sensitivity because thermal noise at the input sets a lower bound to the minimum detectable signal. Consequently there is little use for link gains that put the minimum detectable link signal below the input thermal noise.

It is also useful to examine the impact of increasing modulation efficiency on the link IM-free DR. In (7.9) we presented an expression for third-order intermodulation suppression in which we saw that IM-free DR was inversely proportional to V_π^2. Consequently IM-free DR has the inverse dependence on modulator V_π from that of noise figure. Thus over the range of V_π where both noise figure and IM suppression are proportional to V_π^2, the IM-free DR is independent of V_π. In the region where noise figure is independent of V_π, the IM-free DR increases quadratically with V_π.

Also plotted in Fig. 7.6 is the IM-free DR vs. V_π. For large V_π – i.e. for low incremental modulation efficiency – the intermodulation terms are above the thermal noise floor and the IM-free DR is independent of V_π. Once the intermodulation terms fall below the noise floor, the IM-free DR decreases with further decreases in V_π.

By viewing both noise figure and IM-free DR vs. V_π we see that there is an optimum range of V_π, which for the particular set of link parameters chosen for this example is between about 0.05 and 0.5 V, depending on the CW optical power. Given that present Mach–Zehnder modulators have V_πs in the range of 2–10 V, it is easy to see why there is the general impression that smaller V_π is better – because that is where we are with the present state of the art. However, as we have seen from this discussion, once it is possible to fabricate modulators with sensitivities in the optimum range, further decreases in modulator sensitivity would actually be counterproductive in terms of minimizing noise figure and maximizing IM-free DR.

7.2.3 SNR vs. noise limits and tradeoffs

It is often assumed, usually to facilitate the analysis, that a link is shot-noise limited, in which case the signal-to-noise dynamic range (SNR) can be made arbitrarily large – at least in principle. The shot-noise-limited SNR vs. photodiode current is plotted in Fig. 7.7.

However, in many practical links (virtually all direct modulation links among them) the laser relative intensity noise (RIN) dominates over the shot noise. Thus it is important to understand the SNR limit imposed by the RIN (Cox and Ackerman, 1999).

For intensity modulated links, the residual intensity fluctuations at the photodetector, with no signal applied to the modulation device, clearly will limit the minimum intensity modulated signal that can be conveyed by the link. Consequently it is clear intuitively that the RIN must impose a limit on signal-to-noise ratio. To formalize this limit, consider the link shown in Fig. 5.4. Since RIN-limited detection is being assumed here, that is the only photodiode noise source we consider in this analysis. The photodiode modulation current, i_d, is related to the average detector

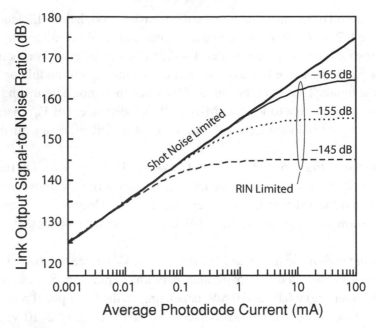

Figure 7.7 Plot of the SNR vs. average photodiode current for shot-noise-limited detection and three cases of RIN-limited detection.

current I_D by the optical modulation depth m, viz.:

$$i_d = mI_D, \tag{7.11}$$

Under the assumption of RIN-limited detection, the link's output SNR can be written as

$$SNR = 10\log\frac{m^2}{rin} = 10\log m^2 - RIN. \tag{7.12}$$

However, recall from (5.10) that one definition of RIN is

$$RIN = 10\log\frac{2\langle i_d^2\rangle}{\langle I_D\rangle^2 \Delta f}, \tag{7.13}$$

where Δf is the photodiode bandwidth. Substituting equation (7.13) into (7.12) yields

$$SNR = 10\log\frac{m^2\langle I_D\rangle^2 \Delta f}{2\langle i_d^2\rangle}. \tag{7.14}$$

In Fig. 7.7 we also plot the RIN-limited SNR vs. the average photodiode current, with RIN as the parameter. These plots make clear that in the range of average photodiode currents presently used in analog links, 1 to 10 mA, the RIN of most present diode lasers will limit the SNR.

The maximum $m = 1$, so the maximum SNR under RIN-limited detection is simply

$$SNR_{max} = 10 \log \frac{\langle I_D \rangle^2 \Delta f}{2 \langle i_d^2 \rangle} = -RIN. \qquad (7.15)$$

As a practical example, consider the CATV application where typically $m = 4\%$. Distributed-feedback (DFB) lasers, whose RIN is usually around -155 dB/Hz, are often used in these applications. This combination of parameters results in a maximum SNR $= 61$ dB for $\Delta f = 4$ MHz, which is also representative of CATV applications.

This analysis makes clear that for RIN-limited detection, the only two ways to improve the SNR are either to reduce the RIN or to increase m. The CATV industry has focused primarily on increasing the SNR by reducing the RIN.

Over the last few years there has been considerable interest in increasing the IM-free DR of optical links. A variety of linearization approaches have been investigated, both theoretically and experimentally, as was discussed in Section 6.4. Improvements of 10 to 20 dB in the IM-free DR have been demonstrated, which is sufficient to meet the needs of the CATV application.

A natural question is: how much more improvement is possible with better linearization schemes? Perfect linearization would result in a linear modulation function, independent of the modulation signal amplitude. Recall from Chapter 6 that the IM-free DR is defined as the highest SNR for which the intermodulation terms are equal to the noise floor. But with perfect linearization, there are no intermodulation terms, so the maximum IM-free DR is the SNR. Although linearization was originally motivated by a need to reduce distortion, as was seen above, linearization also has the additional advantage of permitting a larger modulation depth to be used for a given amount of distortion, thereby enabling the SNR to more nearly approach the RIN-limited value.

To help visualize the SNR limit to the IM-free DR, consider the plot shown in Fig. 7.8. This graphs, on log-log axes, the link RF output powers for the fundamental and third-order distortion outputs as a function of input RF power to the modulation device. The maximum SNR is indicated as the difference in powers between the maximum signal – before saturation sets in – and the noise power. The IM-free DR is indicated as the difference in powers between the fundamental power that results in distortion terms equal to the noise power.

To use the above limits as a gauge against which to judge present linearization methods, consider the reported results for external modulation links. Without linearization, an externally modulated link using a standard Mach–Zehnder modulator

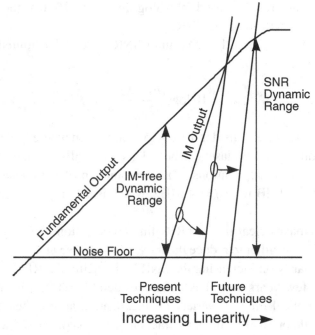

Figure 7.8 Plot of link fundamental and third-order RF output powers vs. link RF input power. (Cox and Ackerman, 1999, Fig. 11. © 1999 International Union of Radio Science, reprinted with permission.)

has an IM-free DR of about 110 dB Hz$^{2/3}$ and a maximum SNR of 160 dB Hz (Betts *et al.*, 1990). Using a solid state laser, the RIN is not directly measurable, but measurements from which RIN can be inferred place an upper bound of -180 dB/Hz. The best reported linearization methods have IM-free DR of 135 dB Hz$^{4/5}$ when used with the same optical power as the link with a standard modulator (Betts, 1994). Thus there is room for an improvement in IM-free DR of about 25 dB over what has been demonstrated to date.

In Section 6.4 both narrow and broad bandwidth linearization methods were discussed. The impression from this discussion was that the primary disadvantage to broad bandwidth linearization was an implementation issue; it required more complex modulators and bias control circuits. However, Bridges and Schaffner (1995) have pointed out that there is a more fundamental reason for preferring narrow bandwidth linearization.

Figure 7.9 is a plot of link noise figure vs. IM-free DR. Each data point shows the combined performance that was achieved for a particular link. The interesting fact to notice is that broad bandwidth linearization – at least as implemented to date – imposes a noise figure penalty, whereas narrow bandwidth linearization does not.

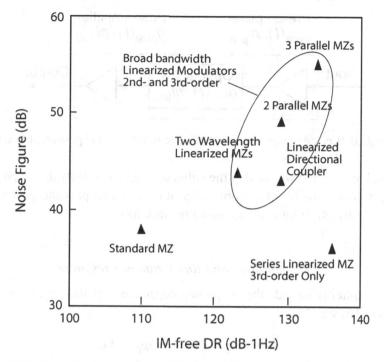

Figure 7.9 Plot of noise figure vs. IM-free DR for external modulation links using narrow and broad bandwidth linearization (after Bridges and Schaffner, 1995).

To date this effect is based entirely on empirical data. However, since it is seen with a variety of modulator types, there likely is some underlying basis for it.

7.3 Tradeoffs between intrinsic link and link with amplifiers

In this section we expand the scope of our analysis beyond the intrinsic link to include electronic pre- and post-amplifiers as shown in Fig. 7.10. Our goal is to derive expressions for the parameters of this subsystem in terms of the link and amplifier parameters.

7.3.1 Amplifiers and link gain

The subsystem gain, g_s, can be written in terms of the pre-amplifier, link and post-amplifier gains, g_{pre}, g_{link}, and g_{post}, respectively, from inspection of Fig. 7.10:

$$g_s = g_{pre}g_{link}g_{post}. \tag{7.16}$$

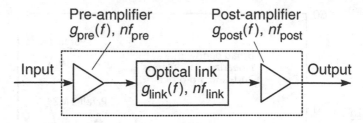

Figure 7.10 Block diagram of a link with electronic pre- and post-amplification.

Taking the log of (7.16) we see that the subsystem gain is simply the algebraic sum of the individual gains in dBs. From (7.16) it is clear that pre- and post-amplifier gains are equally effective at overcoming the link loss.

7.3.2 Amplifiers and link frequency response

If we now explicitly include the frequency dependence of the gain for each component in (7.16) we obtain

$$g_s(f) = g_{pre}(f)g_{link}(f)g_{post}(f). \tag{7.17}$$

From this expression we see that in general the frequency response of one component in a cascade cannot be *increased* by another component in the cascade. Thus if a modulator has essentially no frequency response above 1 GHz – i.e. if g_{link} $(f) \cong 0$ for $f > 1$ GHz, then a pre- or post-amplifier cannot increase the subsystem bandwidth beyond that of the modulator, even if they have gain out to 10 GHz.

The one slight exception to this is a technique called emphasis, which can be either pre- or post-emphasis. If the link frequency response rolls off but does not go to zero, then the pre- or post-amplifier gain vs. frequency function can be chosen to increase such that it just compensates the link roll-off; for example if pre-emphasis is used we would choose

$$g_{pre}(f) = g_{link}^{-1}(f). \tag{7.18}$$

7.3.3 Amplifiers and link noise figure

The subsystem noise figure of the cascade in Fig. 7.10 in terms of the component noise figures is more involved than the gain expression and consequently it is less likely one can write it by inspection. However, cascades of RF components are often encountered in RF design. Thus such an expression is common in noise texts. Although the derivation takes only about a page (see for example Motchenbacher and Connelly, 1993), we simply present the result here:

$$nf_s = nf_{pre} + \frac{(nf_{link} - 1)}{g_{pre}} + \frac{(nf_{post} - 1)}{g_{pre}g_{link}}, \qquad (7.19)$$

where we have assumed that there is impedance matching throughout.

There are a couple of things to note about (7.19). The first is that (7.19) is in terms of noise factor, not noise figure. A second note is that the subsystem noise figure cannot be less than the pre-amplifier noise figure.

A third aspect of (7.19) is that the noise contributed by each succeeding stage in the cascade is attenuated by all the gains that precede that stage. Therefore a low noise, high gain pre-amplifier can be very effective at reducing the effects of a high noise link, whereas a post-amplifier often has no effect. This result is analogous to the effect we saw in link design where the modulation device slope efficiency was more effective at reducing the intrinsic link noise figure than was the photodetector slope efficiency.

A final note on (7.19) is that in principle *any* link noise figure can be made negligible compared to the pre-amplifier noise figure *if* a sufficiently high gain amplifier is available. Given this last statement, one may ask: why did we spend so much time trying to reduce the noise figure of the intrinsic link? There are two answers to this question, one practical and one that involves a design tradeoff. The practical answer is that it is not always possible to obtain a pre-amplifier with the required gain. For example a 20 GHz link with a 50 dB noise figure would require an amplifier that presently does not exist. The tradeoff answer involves the IM-free DR and will be deferred to the next section.

To obtain a feel for how the variables in (7.19) affect the subsystem noise figure, we have prepared the plot shown in Fig. 7.11. We assume that there is no post-amplifier and plot the subsystem noise figure (i.e. pre-amplifier + link) vs. pre-amplifier gain with link noise figure as a parameter. We have assumed that the pre-amplifier noise figure is 1 dB and is independent of its gain. We see that for pre-amplifier gains greater than the link noise figure, the subsystem noise figure is approximately equal to that of the pre-amplifier noise figure.

The conclusion of this section is that electronic pre-amplification is well suited to reducing the high noise figure of present optical links.

7.3.4 Amplifiers and link IM-free dynamic range

As was discussed in Section 6.2.1, distortion arises from the non-linear terms in the power series representation of a device transfer function. Consequently linear amplification – either before or after a non-linear component – cannot re-move the distortion terms generated by that component. Therefore linear pre- or

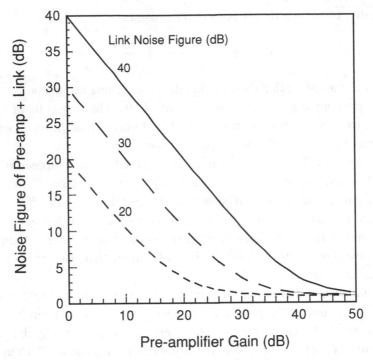

Figure 7.11 Plot of noise figure for the combined pre-amplification + link vs. pre-amplifier gain with link noise figure as the parameter.

post-amplification cannot be used to increase the dynamic range of a link. However, non-linear gain, if properly chosen, potentially can increase the IM-free DR. As was discussed in Section 6.4.1, pre-distortion – which in the context of the present discussion can be viewed as non-linear pre-amplification – can be used to increase the IM-free DR.

On the other hand, amplification can decrease the IM-free DR, and as we will show in this section, the greater the amplification the greater the degradation in IM-free DR. This degradation in IM-free DR introduces a tradeoff with respect to amplifier gain, since as we just saw in the preceding section one generally needs high gain to reduce the noise figure.

To establish a quantitative relationship among the distortion parameters of the intrinsic link, the amplifiers and the link + amplifiers combination, we need a formula analogous to the multi-stage noise figure formula we presented in (7.19). As might be imagined, it does not take a very complex network to have distortion equations that are so complex that they obscure any design insight. To avoid this situation, we begin by making several simplifying assumptions.

One is that the intrinsic link is only augmented by a pre-amplifier, i.e. there is no post-amplifier. This situation is often encountered in practice since, as we saw

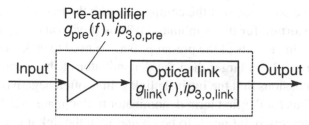

Figure 7.12 Block diagram of link with electronic pre-amplification only.

in Section 7.3.3, only pre-amplification is effective at reducing the noise figure. Thus we only need to consider the two stage cascade shown in Fig. 7.12. Also we assume perfect matching between stages as well as at the pre-amplifier input and link output.

Another assumption is that third-order distortion dominates both the pre-amplifier and the intrinsic link. We further assume that there is only one source of third-order distortion in the link. These assumptions reduce the distortion calculation to one that is often found in RF design: calculating the overall distortion that results from the cascade of two, or more, distortion-producing components. Kanaglekar *et al.* (1988) have derived such an expression; like the analogous noise figure equation, the derivation is straightforward. Hence we will not reproduce it here, but state only the result for the third-order output intercept point, $ip_{3,o,s}$, of the cascade of two distortion producing components in terms of the third-order output intercept points of the pre-amplifier and link, $ip_{3,o,pre}$, $ip_{3,o,link}$.

$$ip_{3,o,s} = \left(\frac{1}{ip_{3,o,pre}^2 g_{link}^2} + \frac{1}{ip_{3,o,link}^2} + \frac{2\cos(\phi_{pre} - \phi_{link} + \theta_{pre})}{ip_{3,o,pre} ip_{3,o,link} g_{link}} \right)^{-\frac{1}{2}}, \quad (7.20)$$

where g_{link} is the intrinsic link gain.

We see from (7.20) that the total distortion from the pre-amplifier + link cascade is simply the sum of two, uncorrelated distortion sources – one in the pre-amplifier and one in the link. These distortion sources are uncorrelated since they arise through completely separate processes. The minimum IM-free DR corresponds to the case where the cosine term in (7.20) is one. Kanaglekar *et al.* (1988) have shown that the minimum IM-free DR is about 1 to 2 dB lower than the expected value of the IM-free DR. Hence this assumption is slightly, but not overly, conservative.

Applying this assumption to (7.20), we recognize that what remains is a perfect square, which cancels out the square root. Thus the expression for the minimum third-order output intercept point of the cascade is simply

$$ip_{3,o,s} = \left(\frac{1}{ip_{3,o,pre} g_{link}} + \frac{1}{ip_{3,o,link}} \right)^{-1}. \quad (7.21)$$

From (7.21) we conclude that the component with the lower intercept point will dominate the distortion for the combination. Thus if we want to minimize the dynamic range limiting effects of the pre-amplifier, we need to make the denominator of the first term in (7.21) larger than the denominator of the second term. Recall from earlier discussions that the intrinsic link gain is often negative in dB, i.e. it is less than one. Thus for the first term denominator to dominate over the second, the pre-amplifier intercept point needs to be *greater* than the link intercept point by at least the intrinsic link gain.

This gives rise to the noise figure vs. dynamic range tradeoff that was mentioned in Section 7.3.3. In that section we saw that the pre-amplifier gain needs to be at least equal to the link noise figure, which in turn has a lower bound set by the intrinsic link loss. Consequently the pre-amplifier intercept point needs to be greater than the intrinsic link intercept point by at least the pre-amplifier gain.

A numerical example will help to underscore the impact of this result. Suppose that a link has an intrinsic gain of -30 dB and a third-order, output intercept point of 0 dBm. Then we require a pre-amplifier with a gain of at least 30 dB and a minimum third-order, output intercept point of 30 dBm, or 1 watt.

To visualize the tradeoffs involved, we have plotted the pre-amplifier + link noise figure and third-order IM-free DR vs. pre-amplifier gain with pre-amplifier IP_3 as a parameter in Fig. 7.13; other link and pre-amplifier parameters are as listed in Table 7.2. Equation (7.19) was used to calculate the noise figure of the pre-amplifier + link combination as a function of the pre-amplifier gain and noise figure; in this somewhat idealized example, it is assumed that these parameters can be obtained independent of the pre-amplifier intercept point. To calculate the IM-free DR of the pre-amplifier + link combination, (7.21) was used to calculate the intercept point of the combination as a function of the pre-amplifier intercept point and (7.4) was used to convert the noise figure and intercept point of the combination into IM-free DR for the combination.

As expected, for pre-amplifier gains on the order of the intrinsic link noise figure, 40 dB, the noise figure of the pre-amplifier + link cascade approaches the pre-amplifier noise figure. With 30 dB of intrinsic link loss and an intrinsic link third-order output intercept point around 0 dBm, this link requires a pre-amplifier with an intercept point of roughly 30 dB + 0 dBm = 30 dBm (1 W) to avoid seriously degrading the dynamic range of the intrinsic link.

Schaffner and Bridges (1993) have extended the above analysis to the case with a linearized link dominated by fifth-order distortion that is preceded by a pre-amplifier dominated by third-order distortion.

In summary of this section on the effects of combining electronic amplifiers with intrinsic links, we have prepared Table 7.3. This table underscores the limited utility of post-amplification, which can only affect the gain. On the other hand,

Table 7.2 *Parameter values used to generate the plots shown in Fig. 7.13*

Pre-amplifier noise figure	1 dB
Intrinsic link gain	−30 dB
Intrinsic link noise figure	39 dB
Intrinsic link third-order IM-free DR	110 dB Hz

Table 7.3 *Summary of the impacts of electronic pre- and post-amplification on the key link parameters*

Parameter	Pre-amplifier	Post-amplifier
Gain	yes	yes
Frequency response	no	no
Noise figure	yes	no
Dynamic range	no	no

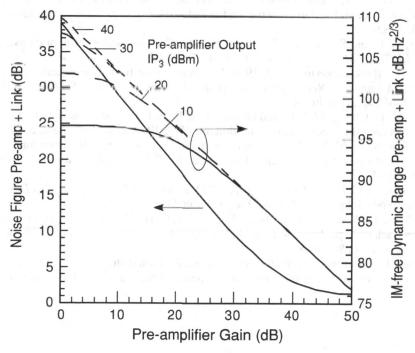

Figure 7.13 Pre-amplifier + link noise figure and third-order IM-free DR vs. pre-amplifier gain with pre-amplifier third-order intercept point as a parameter; other link and pre-amplifier parameters as listed in Table 7.2.

pre-amplification – which can also affect the gain – can be very effective at reducing the intrinsic link noise figure. But as we have seen immediately above, the gain required to improve the noise figure will likely come at the expense of a reduction in dynamic range. Neither pre- nor post-amplification can increase the bandwidth.

References

Ackerman, E. I., Wanuga, S., Kasemset, D., Daryoush, A. S. and Samant, N. R. 1993. Maximum dynamic range operation of a microwave external modulation fiber optic link, *IEEE Trans. Microwave Theory Techn.*, **41**, 1299–306.

Ackerman, E. I. 1998. Personal communication.

Betts, G. 1994. Linearized modulator for suboctave-bandpass optical analog links, *IEEE Trans. Microwave Theory Tech.*, **42**, 2642–9.

Betts, G. E. and O'Donnell, F. J. 1993. Improvements in passive, low-noise-figure optical links, *Proc. Photonics Systems for Antenna Applications Conf. – III*, Monterey, CA.

Betts, G., Johnson, L. and Cox, C., III. 1990. High-dynamic-range, low-noise analog optical links using external modulators: analysis and demonstration, *Proc. SPIE*, **1371**, 252–7.

Betts, G. E., Donnelly, J. P., Walpole, J. N., Groves, S. H., O'Donnell, F. J., Missaggia, L. J., Bailey, R. J. and Napoleone, A. 1997. Semiconductor laser sources for externally modulated microwave analog links, *IEEE Trans. Microwave Theory Tech.*, **45**, 1280–7.

Bridges, W. B. and Schaffner, J. H. 1995. Distortion in linearized electrooptic modulators, *IEEE Trans. Microwave Theory Tech.*, **43**, 2184–97.

Cox, C. H., III and Ackerman, E. I. 1999. Limits on the performance of analog optical links, Chapter 10, *Review of Radio Science 1996–1999*, W. Ross Stone, ed., Oxford: Oxford University Press.

Farwell, M. L., Chang, W. S. C. and Huber, D. R. 1993. Increased linear dynamic range by low biasing the Mach–Zehnder modulator, *IEEE Photon. Technol. Lett.*, **5**, 779–82.

Kanaglekar, N. G., McIntosh, R. E. and Bryant, W. E. 1988. Analysis of two-tone, third-order distortion in cascaded two-ports, *IEEE Trans. Microwave Theory Tech.*, **36**, 701–5.

Motchenbacher, C. D. and Connelly, J. A. 1993. *Low-Noise Electronic System Design*, New York: John Wiley & Sons, Inc., Section 2.8.

Roussell, H. V., Helkey, R. J., Betts, G. E. and Cox, C. H. III 1997. Effect of optical feedback on high-dynamic-range Fabry–Perot laser optical links, *IEEE Photon. Technol. Lett.*, **9**, 106–8.

Schaffner, J. H. and Bridges, W. B. 1993. Intermodulation distortion in high dynamic range microwave fiber-optic links with linearized modulators, *J. Lightwave Technol.*, **11**, 3–6.

Index

applications
  CATV 9–10
  cellular 8, 11
  PCS, *see* cellular
  radar 8, 11
amplifier
  electronic (RF) 8–9, 73, 75
  optical 10–11, 13
  pre-amplifier and link distortion 280–282
  pre-amplifier and link noise figure
      278–279
  pre-amplifier noise figure vs. IM-free DR,
      tradeoff 282–284
available (modulation) power, definition 27

bandwidth, *see* modulation bandwidth
Bode–Fano limit; *see* intrinsic link gain, gain vs.
      bandwidth tradeoff

compression, *see* intrinsic optical link gain,
      compression

detection
  coherent 5
  direct 5
detector, *see* photodiode
diode laser, *see* laser
direct modulation, *see* specific aspect of, e.g. gain,
      noise figure, etc.
distortion
  clipping (large signal) 204, 218
  composite second order (CSO)
      212–215
  composite triple beat (CTB) 212–215
  definition of 202
  harmonic 205–206
  intercept point 209–210
  intermodulation 205–206
      suppression of 216
  laser
      as function of bias 264
      low frequency 217–220
      high frequency 220–222

modulator
  directional coupler 225–227
  electro-absorption 227–228
  Mach–Zehnder 222–224
      as function of bias 270–271
      pre-amplifier and link 280–282
  order 207, 208–209
  photodiode 228–232
  relationship between power and voltage or current
      204

external modulation, *see specific aspect of, e.g. gain,
      noise figure, etc.*

feedback, *see* linearization technique feedback
feedforward, *see* linearization technique feedforward
fiber, *see* optical fiber
frequency response, *see* modulation bandwidth

impedance matched link examples
  direct modulation 127–130
  external modulation 136–139
impedance matching
  conjugate match condition 111
  diode laser
      lossless conjugate match 124
      lossless magnitude match 119
      lossy match 118
  lossless vs. lossy match 111
  Mach–Zehnder modulator
      lossless conjugate match 134
      lossless magnitude match 132
      lossy match 130–132
  magnitude match condition 111
  PIN photodiode
      lossless conjugate match 115–116
      lossless magnitude match 112
      lossy magnitude match 112–113
incremental detection efficiency, PIN 52–54, 71
      lossless conjugate match 116–117
      lossless magnitude match 113–114
          gain/bandwidth tradeoff 114–115
      transformer matched 179

incremental modulation efficiency 71
  diode laser 30
    lossy magnitude match 118–119
    lossless magnitude match 120–121
      gain/bandwidth tradeoff 121–124
    lossless conjugate match 124–125
      gain/bandwidth tradeoff 126
  directional coupler modulator 43–45
  electro-absorption (EA) modulator 48
  Mach–Zehnder modulator 37–39
    for arbitrary bias point 269, 271–273
    lossy magnitude match 131–132
    lossless magnitude match 132–133
      gain/bandwidth tradeoff 133–134
    lossless conjugate match 134–136
    transformer matched 179
intercept point
  definition, *see* distortion
  of preamplifier and link 280–282
intermodulation-free dynamic range 210–211
  as function of bias
    diode laser 264
    Mach–Zehnder modulator 270–271
  relation to composite triple beat (CTB)
    216–217
  relation to intercept point 211
  scaling with bandwidth 211
intermodulation suppression 216
intrinsic optical link gain 71, 91
  compression (or gain compression) 82
    definition 1 dB 83
    diode laser 83
    Mach–Zehnder 84–87
  cross-over optical power 78
  definition 71
  direct modulation 72–73
  scaling with
    optical power 76
    slope efficiency and responsivity 81–82
    wavelength 79
    positive 80
      via impedance matching 148–150
  efficiency 79, 87
  external modulation 74–75
  as function of Mach–Zehnder bias point 269
    positive 74, 77–78, 103, 138
    scaling with
      optical power 76–77
      slope efficiency and responsivity 81–82
      wavelength 80–81
  gain vs. bandwidth tradeoff
    lossy impedance matching 139–142
    lossless impedance matching (Bode–Fano Limit)
      comparison with experimental results 150–151
      diode laser example 144–145
      direct modulation link example 145–149
      integral form 143
      with most effective form of reflection
        coefficient 143
  impacts of optical loss on 72
  transformer matched 180

laser
  basic operation 21–23
  distributed feedback (DFB) 31–32, 76
  Fabry–Perot 21–24, 76–77, 121
  facet (mirror) reflectivities 24, 31, 32
  in-plane 27, 32, 72, 91, 118
  longitudinal mode spacing 23
  gain-guided 24
  index-guided 24
  power vs. current (P vs. I) 24–25
  quantum efficiency, external differential 26
  rate equations 24
  slope efficiency, *see* slope efficiency, laser
  spontaneous emission 22, 24
  stimulated emission 22, 95
  threshold current 25, 95
  vertical cavity, surface emitting (VCSEL)
    32–34
  wavelength (free space) 23
  wavelength range 13
large signal modulation efficiency 85
linearization techniques
  dual series Mach–Zehnder
    narrow bandwidth (sub-octave) 243–247
    wide bandwidth 241–244
  feedback 234–235
  feed-forward 235–236
    power requirements 235–236
  frequency range 247–249
  hybrid 236–237
  optical 240–243
  pre-distortion 233–234
    bandwidth required 233–234
    example 237–240
link, *see* optical link

modulation 4–5
  amplitude 5, 6
  direct 6, 8, 10, 11, 20, 96
  external 6, 8, 10
  indirect, *see* external
  intensity 6
modulation bandwidth
  3 dB optical vs. electrical 92–94
  diode laser 94, 96
    as function of bias 264
    modulation above relaxation resonance 98
    relationship to slope efficiency 97
    relaxation resonance 94–96
  directional coupler 98
  electro-absorption, *see* photodiode
  link, in terms of device bandwidths 126–127
  Mach–Zehnder modulator 98
    antenna coupled 104–105
    lumped element 99
    phase reversal 101
    relationship to slope efficiency 103–104
    traveling wave 100
      electrode length-bandwidth product 100
      factors limiting maximum length 101–103
      velocity matched 101

photodiode and EA modulator 105
  bandwidth-responsivity tradeoffs 106–110
  surface illuminated 107–109
  traveling wave 109–110
  trade off with link gain, *see* intrinsic link gain, gain vs. bandwidth
modulation depth (or index) 205
  impact on SNR and distortion 207–208
modulation device 2, 3, 6, 69
  transfer function 205
  wavelength range 13
modulator
  directional coupler 41
    transfer function 41–44
      compared to Mach–Zehnder 44, 45
      V_s 41
  electro-absorption (EA) 45
    Franz–Keldysh effect (FKE) 46–47
    quantum confined Stark effect (QCSE) 46
    transfer function 47–48
    V_α 48
  Mach–Zehnder 35–36, 74, 91, 129, 202
    transfer function 36–37, 84
    V_π 36
  materials, comparison 34, 36

network
  active 186–187
  lossless 185
  passive 185
noise
  addition of two or more sources 168
  equivalents
    RIN equals shot 166
    shot equals thermal 163
  relative intensity noise (RIN)
    definition of 163
    single vs. multimode laser 165–166
    spectrum of 164–165
  shot
    definition of 162
    spectrum of 162–163
  thermal (Johnson)
    definition of 160
    spectrum of 161
noise factor, definition of 167
noise figure, definition of 167
  and linearized IM-free DR 276–277
  as function of bias point
    diode laser 264
    Mach–Zehnder 270
  effects of
    device slope efficiency on
      RIN dominated link 181–182
      shot noise dominated link 180–181
    impedance mismatch on 194–196
    matching circuit on 172, 177, 180
    optical power on
      direct modulation 182–183
      external modulation 183–184
    type of impedance match on 179

frequency dependence 168
limits on
  3 dB 186
  general passive match 191
  lossless
    passive 185–186
    passive match 186
  lossy
    passive 194–196
    passive match 189–194
    passive attenuation 187–188
  of preamplifier and link 278–279
  RIN dominated, direct modulation link 170–174
  shot-noise dominated, external modulation link 173–178, 192–194
notation, for variables and subscripts 20

optical amplifier, *see* amplifier, optical
optical fiber
  attenuation, *see* loss
  dispersion
    chromatic (wavelength) 13, 30
    modal 15
  loss, general 1, 3, 9, 13
    absorption 13
    microbending 16
    scattering (Rayleigh) 13
  type
    multimode 15
    polarization maintaining 16
    single mode 14
optical link 69, 201
  comparison with RF/coax 3–4
  definition 1
  design tradeoffs
    direct modulation
      laser bias, effects of 263–264
      noise figure vs. IM-free DR 265–267
      optical reflections, effects of 267–268
    external modulation
      IM-free DR vs. bias 270–271
      modulator bias point, effects of 268–271
      noise figure vs. RIN and bias 270
      NF and IM-free DR vs. V_π 271–273
  gain, definition 70
  loss
    optical 1, 10
    RF 2, 3, 8, 10

photodetection device 2, 3, 6, 69
  wavelength range 13
photodetector, *see* photodiode
photodiode
  I-layer 49–51
  PIN photodiode 49–50, 72, 74, 76, 91, 112
    waveguide 52–53, 106, 109–110
  responsivity 51–52, 72–73, 74–75, 76
    maximum value 52–53
    wavelength dependence 51–52, 79, 80

pre-amplifier, *see* amplifier
pre-distortion, *see* linearization techniques,
    pre-distortion

quantum efficiency, external differential
    laser 26
    photodiode 51

reflection coefficient
    definition 140, 143
    laser 144
    photodiode 147
relaxation resonance, *see* modulation bandwidth,
    diode laser
resistive match, *see* impedance matching,
    lossy
responsibility, *see* photodiode, responsivity
root mean square (RMS), definition of
    159

signal-to-noise ratio (SNR)
    RIN and shot noise limits 273–275
    as bound on IM-free DR 275–276
slope efficiency
    laser 25–27, 72–73, 76
        as function of bias 263
        maximum value 26
        wavelength dependence 27, 79
        fiber-coupled 27
    modulator
        directional coupler 45
        electro-absorption 48
        Mach–Zehnder 39–41, 74–75, 85
        maximum value 39
        wavelength dependence 40, 80
superposition 170

Taylor series 204
transducer power gain, *see* optical link gain, definition